Enhancing Competitiveness in Small Developing States

Approaches, Tools and Policies

Robert A Stoddard

authorHOUSE

AuthorHouse™ UK
1663 Liberty Drive
Bloomington, IN 47403 USA
www.authorhouse.co.uk
Phone: UK TFN: 0800 0148641 (Toll Free inside the UK)
UK Local: (02) 0369 56322 (+44 20 3695 6322 from outside the UK)

© 2024 Robert A Stoddard. All rights reserved.

No part of this book may be reproduced, stored in a retrieval system, or transmitted by any means without the written permission of the author.

Published by AuthorHouse 08/14/2024

ISBN: 979-8-8230-8754-4 (sc)
ISBN: 979-8-8230-8755-1 (e)

Library of Congress Control Number: 2024908923

Print information available on the last page.

Any people depicted in stock imagery provided by Getty Images are models, and such images are being used for illustrative purposes only.
Certain stock imagery © Getty Images.

This book is printed on acid-free paper.

Because of the dynamic nature of the Internet, any web addresses or links contained in this book may have changed since publication and may no longer be valid. The views expressed in this work are solely those of the author and do not necessarily reflect the views of the publisher, and the publisher hereby disclaims any responsibility for them.

Contents

Preface..xi

1 Introduction..1
   Development Agenda for Small Developing States 5
   Least Developed Countries in a Globalized World....................9
   Development Assistance for Small Developing States 13

2 The Conceptual Notion of "Small States" 18
   African Small Island Developing States...................................23
   Asia, Pacific Island Countries (PICs) 25
   The Smaller Island States (SIS)..27
   Structural Development Challenges for Small States...............28

3 A Regional Manufacturing Sector in Perspective................32
   An Analysis of Fiscal Incentives Provided................................36
   Selected Incentives Provided by Some Small States 37
   Tax Holidays 39
   Specialised Arrangements for Business Development..............42
   Special Zones 43
   Industrial Parks 45
   Cyber Parks 47

4 Technology Transfer Towards Competitiveness 50
   Technology Transfer; A Development Tool.............................51
   Building Technological Capability..54
   Scope and Dimensions of Technology Transfer......................55
   Technology Embodiment Scenario...57
   Technology Transfer Modalities..58
   Non-Commercial Channel of Technology Transfer................64
   Enabling Factors for Technology Transfer 64

    Technology Appropriation..67
    The Impact of Technology Transfer on Development..............68
    Establishing the Framework for the Transfer of Technology.....71

5   Commercial & Technological Innovation........................... 75

    The Innovation Process 76
    Strategic Approach to Innovation 81
    Managing the Innovation Process..83
    Innovation and Marketing in Product Development 85
    Marketing; A Driver of Product Innovation 88
    How Product Design Reflects Consumer Needs 89
    How Marketing Information Enhances Innovative Design.......92
    The Role of Strategic Planning in Marketing Innovation..........93
    Science & Technology and the Evolution of Product
    Development...95

6   Traditional Development Framework for Enterprises 98

    Clustering / Networking in Small Enterprises.........................101
    Business Incubation; Challenges and Opportunities 106
    Capacity Building within Private Sector Organizations..........113
    E-Commerce; A Tool for Competitiveness 115
    E-Commerce Infrastructure..118
    Constraints within the E-Commerce Sector 120
    The Role of E-commerce in Modern Business........................121
    ICT, A Business Facilitating Tool..124

7   World-Class Manufacturing & Management
    Techniques 127

    The Concept of World-Class Manufacturing..........................128
    Evolution of World-Class Manufacturing Techniques...........133
    Features of World Class Manufacturing..................................134
    World-Class Information Systems 135
    World Class Manufacturing Strategy 136
    World-Class Management Techniques and Philosophy..........137

Production Planning and Control Techniques 140
Concurrent Engineering.. 140
Continuous Improvement (Kaizen) 141
Lean Production... 142
Total Productive Maintenance (TPM) 144

8 Managing Quality to Improve Competitiveness 149
Quality Assurance and Quality Control Approach 151
Quality Circles 154
Statistical Process Control (SPC) 155
Total Quality Management (TQM) 156
Total Quality Control 157
Total Employee Involvement 158
Benchmarking as a Competitive Tool 160

9 Standards; A Catalyst for Competitiveness 164
Standards and Quality Approach to Competitiveness 165
How Standards Impact Quality 170
Globalization and Standards 172

10 The Global Value Chain in Perspective............................... 175
Challenges & Opportunities of the Global Value Chain 177
Functioning within the Value Chain Systems 181
The Agriculture Sector and its Linkages to the
Supply Chain 183
Value Chain Challenges and Impact on Key Sectors 187

11 A Strategic Framework to Support SMEs 193
Strategic Continuous Improvement.. 193
Assessment Phase.. 194
The Planning Phase.. 195
Implementation Phase.. 197
Productivity and Productivity Improvement.......................... 199

    Benchmarking within the SCIM Process 203
    Organizational Assessment 204

12 The Paradigm of Competitiveness and Industry 207
    Privatization Option for Public Sector Enterprises................. 209
    Regional Industries & International Competitiveness........... 211
    High-End Value Added 221
    Niche Manufacturing in Small Island Economies 224
    Niche Manufacturing; Challenges and Opportunities 225

13 Development Strategies and their Relevance to SIDS 232
    Comparative Experiences; The Newly Industrializing
    Economies.. 234
    Lessons from Comparative Analysis...................................... 237
    Structural and Economic Constraints 242
    Complementarity 245

14 Industrial Policy Issues; Perspectives and Proposals 251
    Industrial Policy Structure and Framework.......................... 254
    Macroeconomic Policies 259
    Sectoral Policy 261
    Traditional Approach to Industrial Policy 262
    ISI Policy and Strategy Consideration 262
    Industrial Policy Priorities for Small Economies 264
    Strategic Goals of an Effective Industrial Policy 265
    Sectoral Policy Shift towards Sustainable Growth 267
    Policy Constraints within the SME Sector............................ 269
    Human Resource and Industry-led Growth......................... 272
    Technology Transfer Policy Imperatives 276
    R&D a Catalyst for Product Development.......................... 279
    Intellectual Property (IP) 280
    Business Associations... 284
    Infrastructure Development; A Key Prerequisite.................... 286
    Financing and Incentives Requirements 288

Policy Alternative to Incentives..289
Sea and Air Transportation 292
Public-Private Sector Partnership..294
Energy for Sustainable Development......................................299
Industrial Capacity Building...302
Institutional and Legal Strengthening.....................................304
Production Integration 307
Marketing and Distribution Policy...308
Labour Relations and Good Governance...............................309
Environmentally Sustainable Manufacturing.........................312

References 319
Index..339

Preface

During the last two decades, many small developing states have been experiencing significant challenges amidst changing trade rules that have denied them the opportunity of entering and competing in global markets. These factors have had a crippling effect on small enterprises and have retarded their growth and their ability to contribute significantly to economic development. Furthermore, the export trade in some traditional commodities has been severely affected due to a decrease in demand for these products. In some cases, some products have been replaced by low-cost substitutes and synthetic materials. This has resulted in the collapse of trade in some markets.

Most small developing states are characterized by small size, limited resources and lack of access to markets. These factors have stymied their growth for many years. This is evident by a rapid rise in their debt burden which provides them with little fiscal space. For the most part, governments depend on the contribution from the private sector and in particular small enterprises in their quest to provide employment and reduce poverty. In many cases, those enterprises that survive this economic malaise are rife with poor working conditions, low wage offerings and a lack of capital for further investment. The policy and regulatory environment to which they are subjected are more often than not archaic and need to be reformed to align them to current industrial best practices and trends.

To diversify their economies, many of these states turn to the services and the tourism industries for economic growth. The services sector is a nascent sector that requires heavy investment in infrastructure, trained personnel and other technological resources supported by the relevant regulatory framework. Over the years, there has been a tendency among some developing states to play "copycats" and copy many initiatives from developed countries without understanding the cultural and economic dynamics that have been attributed to their success. It requires a paradigm shift in small states that will compel them to develop their own economic model considering their comparative strengths and understanding their cultural heterogeneity, indigenous endowment and factors that impact economic sustainability. Understanding the comparative development path of other small economies with similar challenges and history is useful because there may be important lessons to be learned, but not necessarily to replicate wholesale.

This book addresses many of the existential challenges that have beset small developing states for many years that have become obstacles to their growth and success. It has also proposed several options for industrial and economic development and social transformation. Many initiatives and systems that are associated with new development buzzwords have been discussed and critiqued. Furthermore, alternative recommendations have been proffered based on rational arguments and competitive norms.

Several success stories from the Caribbean and elsewhere have been highlighted as well as anecdotal evidence where costly mistakes have been made with the appropriate caution to avoid them. Some of these cases have been linked to industrial policy issues that have been discussed at length.

Chapter 1

Introduction

Liberalisation of international trade has been in effect for most of the post-war period. However, there were key changes to bolster global trade in the mid-1980s with the launch of the Uruguay Round and in the early 1990s with its successful completion and subsequent creation of the WTO. This new thrust in world trade has resulted in numerous challenges and to some extent opportunities for small developing states, which, for many years, had been the beneficiaries of trade preferences under agreements such as the Lomé Convention and the Cotonou Agreement.

It was envisaged that in the years ahead, economic growth would spiral due to "globalization and liberalization" in trade, the two buzzwords of the 1990s. Although not a new phenomenon, there is no universally accepted definition for the term 'globalization.' It connotes the notion of a process where national markets are becoming increasingly interlinked. It has also enabled the internationalization of production through international trade and capital flows manifested by private firms operating in the global environment. Emanating from the process of globalization are profound changes in technology, new methods in industrial organization and the international division of labour.

As a result of these new trends, global production systems are fragmenting and many activities are being relocated to different parts of the world, while large companies subcontract production to specialised SMEs. It was expected that globalization and liberalisation would to a large extent reduce the divergence between rich and poor countries. These expectations were not fully realized and as a result, many developing countries have struggled since. Several studies have shown that there has been little improvement in real income and economic growth has been slow and unstable. The situation has led to a rethinking of development policies in low income countries that have led to the consensus that poverty reduction should be a key focus among developing countries, particularly small developing states.

Against the background of a growing fragile industrial climate, many small developing states are confronted by a plethora of global scenarios to which they must respond with alacrity by initiating steps to participate in activities from which they can benefit. To confront the challenges posed by an increasingly globalized economy, many countries opt for regional integration where groupings of countries can devise strategies to cope with the challenges of a globalized economy. In this context, regionalism particularly for small developing countries must take precedence over globalization.

Regionalism can embolden countries within specific groupings to develop their competitiveness within regional development schemes. The contrary will expose individual countries to the vagaries of globalization with negative consequences.

Globalization has engendered a pattern of cross-border activities where firms can get involved in bilateral agreements leading to international investment, trade and cooperation for purposes of product development, production, sourcing and marketing.

Some developing states, particularly those in the Pacific and the Caribbean are geographically positioned and politically committed to regionalism because they perceive economic gains can be achieved through economic integration of one form or another, which can have a positive impact on the industrial sector. Hemispherically, the Free Trade Area of the Americas (FTAA), which was established in 2005, was designed to liberate trade in goods within the hemisphere by progressively removing tariff barriers to zero tariff level. Among the potential benefits to small developing states within the Caribbean region, for instance, is enhanced access to the United States markets. In the case of the Pacific Island Countries (PIC), the proximity to their main market of Australia and New Zealand is a distinct advantage. Although some countries are competitive in terms of services, they will need to develop competitiveness in other sectors to derive net benefits in an enlarged free trade zone. Regionalism seems to offer some hope for small developing states since their combined efforts can redound to some form of economic benefit.

Many small states form regional integration agreements that allow preferential trade access to their member states, but at the same time, they maintain an unaltered external trade policy toward the rest of the world, thereby resulting in reduced profits for the entire bloc. Most of these states with similar geopolitical characteristics band themselves together in small trading and economic blocs, although relative geographical dispersion has made regional integration and harmonization a slow process.

Economic integration is becoming an integral part of regionalism as in the case of the Caribbean Community (CARICOM), which has fifteen member states, the sub-regional group of the Organization of Eastern Caribbean States (OECS) has seven member states, and the Pacific Islands Forum which represents heads of government of all sixteen independent and self-governing Pacific Island Countries, Australia, and New Zealand that subsumes the Smaller Island States (SIS) of eight small states.

In the meantime, arrangements are being pursued to deepen the economic integration movement within the CARICOM community towards a single market and economy. This arrangement will facilitate the free movement of goods, capital and labour, thereby overcoming the barriers of time, space, and location and facilitating trade and investment. These factors augur well for the industrial sector in the Caribbean region, in that, it will provide protective space which will enable industries in the small developing states that are part of the community to develop and improve their competitiveness through production efficiency and effective utilization of the factors of production.

For many decades, small developing states like those that are endowed with indigenous agricultural commodities (raw materials) exported them primarily in their primary or semi-finished state. This was a strategic colonial arrangement that provided Britain and other colonial powers with the raw materials required to drive their factories and establish their trading dominance. This pattern of trade in goods continued for many decades and these small states often received low prices in the regional and international markets for these commodities. Because of their heavy reliance on agricultural and extractive industries, they have seen the protective policies disappear and their products commoditized.

Over the last couple of decades, many small states have witnessed the erosion of their trade preferences; a rapid rise in their debt burden; increased susceptibility to climate change and rising concerns about security and crime. All these factors have a significant impact on development. Consequently, some states have turned to manufacturing by which they convert raw materials to semi-finished or finished products that will fetch a high price on the export market. This has increased export trade from developing countries to developed countries.

In subsequent years, some small developing states have experienced serious challenges in manufacturing export. As competitiveness in manufacturing declined in the 1990s, the advent of liberalised trade forced many developed countries to remove the barriers to trade employed earlier as a means of encouraging free trade among countries. For instance, in 1991, the Common External Tariff (CET) instituted by the Caribbean Community (CARICOM) was reduced from between 0-70 percent to between 0-45 percent and a further reduction to 0-20 percent five years later. In addition, the General Agreement on Tariffs and Trade (GATT) and its successor agreement WTO, which was established on January 1, 1995, also accelerated the process of opening the markets of all economies.

Most countries are signatories to the WTO agreement, including small developing island states. As such, they have effectively committed themselves to the fulfillment of the objectives of achieving global free trade. Hence, they are obliged to comply with the organization's rules governing international trade by way of tariffs and quotas imposed by industrialised countries and by constraints on the supply side, due to a lack of technology, skills, training and capital. The advent of the WTO ushered in a new paradigm in world trade that promotes trade liberalisation by strongly advocating the removal of trade barriers in enforcing its policy of open competition. Against the foregoing, enterprises in small developing states must compete by satisfying those requirements or exit the market.

Development Agenda for Small Developing States

Notwithstanding the enormous benefits of trade liberalisation, it has given rise to an increase in imports among third world countries and to some extent has put pressure on local industries, thereby restricting the growth of regional trade. Over the last four decades, particularly following the oil crisis of 1973, several industrialised countries that were former colonial powers have been seeking special dispensation in the international trading system for fairer and more stable prices for their raw materials as well as market access for their industrial products. It was in their best interest to do so because it guaranteed them a sustainable supply of raw materials. Consequently, eight proposals were forwarded by most of the developing countries. Among other things, these proposals include improving the terms of trade, increasing development assistance and developing country tariff reductions. The effect of such lobbying efforts resulted in the declaration of a New International Economic Order (NIEO) in May of 1974, at a special session of the United Nations General Assembly through the United Nations Conference on Trade and Development.

Many of the initiatives proposed by the United Nations to assist struggling developing nations have not yielded the success that was anticipated, evidenced by the fact that there has been little improvement in real income and economic growth in these countries. This reality has led to a rethink of development cooperation for struggling developing countries by multi-national agencies such as the World Bank and the IMF. Consequently, there was a new consensus among the international community that poverty reduction should be the new focus. Resulting from this new thinking was a reclassification of the economic status of developing states. Hence, countries that were hitherto considered poor have now been classified as middle-income states.

It is interesting to note, that leading organizations such as the World Bank and United Nations Industrial Development Organization (UNIDO) that have been involved in assisting disadvantaged groups of countries around the world have identified industrial development and in particular manufacturing as the main engine of growth for the economic survival of small states listed under the category of LDCs. The statistics bear evidence of the fact that those countries with a fragile manufacturing base have low value-added industries and are often doomed to economic failure. No one solution can be applied to arrest this economic situation. There is, however, a range of macroeconomic strategies and technological approaches that can be considered strategic interventions to enhance the well-being of small states.

There is an abundance of evidence to support the notion that industrialization is a fundamental requirement and a central and strategic pillar of economic development. Not only is industrialization the normal route to national development, but because of the globalization of industry, it can also accelerate the pace of development. The potential for accelerated growth could ideally be triggered by manufacturing with all the right economic conditions. It can be a decisive strategy to create opportunities for the poor and marginalised to break out of poverty and for middle-income countries to move up the economic ladder.

Small island economies are experiencing many challenges because of the effects of international economic conditions. Because these island states constitute small economies, the external trends and shocks create adverse situations for promoting growth and stability in their respective regions. Many of the challenges arise from fiscal imbalances and economic growth which fundamentally require countercyclical measures to sustain the economy. In the absence of a strategy to foster private sector development by increasing competitiveness and attracting foreign investment, sustained economic growth will not be possible. To face the economic challenges and improve the competitiveness of any region, the private sector will need to strengthen inter-firm collaboration, reduce costs while maintaining employment and lay the groundwork for strategies of diversification innovation and value-added exports in the medium term.

Countries do not necessarily create wealth by producing the same goods and services more productively. Change becomes essential in what is being produced and the nature of the production process. Structural transformation is the process whereby countries move to new economic activities that are more productive and are thereby able to pay higher wages. Countries that can upgrade their exports by developing new economic activities tend to grow faster. In recent years, trade, economic integration, competitiveness and private sector development issues have become thematic priorities in the discussion on competitiveness and economic development. Although regional organizations, donor agencies and governments have identified similar strategic priorities, there is no coherent strategy being implemented to forge the agenda forward. Consequently, there is a significant gap between diagnosing the real problem and implementing strategies to address it.

The involvement of the private sector is critical to national development. Hence, appropriate policies and strategies are required to expand the private sector's participation in national development. Such policies should be designed to enable the government to overcome or prevent market failures, reduce burdensome transaction costs, provide an overall sound environment for private sector growth and support public-private participation in matters of competitiveness.

Many small developing states continue to strive for new energy sources, efficient methods of food production, better quality products, improved human health and options for institutional changes and environmentally benign technologies. There are a plethora of studies highlighting the importance of agro-industrial development to the expansion and diversification of the agricultural sector. Many agro-processing operations including jams, jellies, fruit nectars and other beverages have been well established in some regions. The technology which is utilised in the small-scale processing sector for the processing of these products remained traditional and labour intensive. For some agricultural inputs such as agrochemicals and agricultural machinery, most small island states are almost completely dependent on importation from a foreign country.

Adopting world-class manufacturing techniques and maintaining a vibrant agro-industrial sector could invariably result in the expansion of markets for primary agricultural products. This approach can be done through value-added activities and by vertically integrating primary production and food processing systems to minimise post-harvest losses. With a few exceptions, the agro-industrial sector remains rudimentary, underdeveloped and largely without significant institutional, technical and financial support.

For many years, industrial development has been an important driver of economic growth. This is especially true in the case of developing countries and economies in transition. Economic growth has enabled countries to increase their factor productivity, which in turn has established the foundation for improvement in the standard of living. There are, however, other sectors in the economy that depend on strong industrial development for their momentum and success. For instance, an increase in agricultural productivity and the processing of agricultural goods. Many high-tech firms survive and grow by responding to the demands created by manufacturing activities. It must be iterated that the industrial revolution in England and its subsequent spread to countries in Europe and North America became the catalyst for economic growth.

There is an ongoing debate, which suggests that society has entered an era of post-industrialization. Although de-industrialization is progressing at a rather rapid pace, industry is still a dominant force in economic development. Over the last several decades, the services sector, which has become one of the fastest-growing sectors, owes its rapid growth to the corresponding growth in manufacturing output. The growth in telecommunication services could not be possible without continuous demand for manufactured hardware.

In most developing countries today, industrial development continues to be the catalyst for economic growth driven by technological progress which has led to income generation and the creation of employment. Much of the slowdown in manufacturing in developing countries is related to the diversion of manufacturing activities through foreign direct investments and the globalization of production. In small developing economies industrial development contributes significantly to value-added and employment. It is noticeable that for countries in transition, large inefficient state-owned enterprises have given or are giving way to smaller and more efficient private sector entities.

In recent decades, the process of industrialization has undergone profound changes while industrial development has remained the fundamental instrument of economic growth and social development, particularly in developing countries. These developments have been influenced by decisions in the international sphere that were responsible for the removal of natural and legal protection previously granted to many small island states that were once colonies of several developed countries in Europe. Consequently, small states have been exposed to high levels of external competition in an increasingly integrated international market.

Least Developed Countries in a Globalized World

Countries around the world have been categorized into several economic groupings. Among them are More Developed Countries (MDCs); Least Developed Countries (LDCs); Newly Industrialised Countries (NICs) and Small Island Developing States (SIDS). These categorizations are based on several criteria derived from the indices of economic indicators.

Although frequent reference will be made to small island developing states, the status of those countries generally considered as least developed cannot be ignored. The list of Least Developed Countries (LDCs) was approved by the UN General Assembly in 1971, in recognition of the existence of a category of countries whose distinction lies not only in the extent of their overall poverty and unemployment but also in their limited economic, institutional and human resources, often compounded by geophysical challenges.

Small states are characterized by the small size of their economies, which prevents the exploitation of economies of scale. These countries are also typically remote from major markets and are more vulnerable to environmental disasters. There are 10 SIDS among the LDCs. In 2010, the sub-Saharan countries represented 61 percent of the LDCs' population. The largest LDC is Bangladesh, which accounts for 19 percent of the LDCs' total population, followed by Ethiopia (10 %) and the Democratic Republic of the Congo (8 %).

A UNIDO analysis suggests that several LDCs, particularly in Asia, are on a sustained path towards industrialization, while LDCs (mainly in Africa) are facing deindustrialization. Regarding international trade, LDCs continue to play a minor role in world trade despite their remarkable export performance in recent years. Their exports are dominated by resource-based and low technology products and they concentrate on a limited number of markets and products, thereby increasing their vulnerability to external shocks. GDP per capita, which measures the level of total economic output in a year per unit of population, has remained relatively low in most LDCs. In 2009, GDP per capita in LDCs was 4 to 9 times lower respectively than that of MDCs and the world average. Low manufacturing value added (MVA) levels combined with fast-growing populations may explain the low level of MVA per capita in LDCs. LDCs were unable to add value to their exports because they continued to export low-technology goods.

LDCs have a relatively high reliance on the primary sector, which increases their vulnerability to external shocks due, for example, to volatile commodity prices. To build more resilient economies, broadening and deepening the manufacturing sector may be deemed a desirable path as manufacturing brings several potential benefits. Firstly, a vibrant manufacturing sector stimulates technological change by adopting, mastering and developing improved production processes and new technologies, boosting productivity throughout the economy. This will invariably require further development of skills and learning, thereby shifting employment towards highly skilled and better-paid job categories.

Secondly, manufacturing promotes economic growth through forward and backward linkages. Linkages are created when a sector utilizes the products of other sectors as inputs (backward linkages), while other sectors use their output as inputs in their production processes (forward linkages); as a result, the growth of one sector can fuel the development of other related sectors. In addition, increased activities in manufacturing can lead to the development of support sectors such as finance and transport.

Finally, by taking advantage of the globalization of production, LDCs can integrate global production networks and access international markets. The expansion of trade is a central pillar of the globalization of the world economy, with manufactured exports consistently accounting for over 80 percent of total exports since 1990. Compared to commodities, manufactured products are less exposed to price fluctuations and can yield more revenues on account of higher value-added. Shifting economic production towards manufacturing may therefore hold higher promise for long-term growth. Yet, it will be important to lean on the primary sector as a provider of both financial resources that may be needed for investments in the manufacture of intermediary inputs (forward linkages) that can be used in manufacturing production.

Opportunities stemming from the rapidly changing global industrial landscape provide no guarantees for economic prosperity. Formulating the responses needed to convert opportunities into sources of wealth creation is a process that requires both the private and public sectors. The recently published report by the World Bank on the independent Commission on Growth and Development argues that the government and the private sector have critical roles to play in boosting growth in developing countries. In the past, economists believed that the developing world was replete with market failures, which refers to the inability of the market system to provide some necessary goods and services or to provide them in optimum quantities or levels. The only way in which poor countries could escape the poverty traps was through forceful government intervention.

Import substitution policy, planning, and state ownership have produced some success in the past, but where there was little or no attempt to change such policies in response to the new world order (particularly so, when other countries were adopting alternative policies that were proven successful) they have led to colossal failures and crises.

Economic liberalisation and opening up of markets offered many benefits to countries with a focus on export markets.

Furthermore, emphasis on financial interests and the development of skilled workers for new industries resulted in economy-wide growth rates (in output and productivity). It is misleading to believe that state planning and public investment can exclusively act as the driving forces of economic development.

It is increasingly being recognized that developing societies need to integrate private initiatives into a framework of public policy that encourages restructuring, diversification and technological dynamism beyond what market forces on their own would generate. Perhaps not surprisingly, this recognition is now particularly evident in those parts of the world where market-oriented reforms were taken the farthest and the disappointment with the outcome is correspondingly the greatest. Market forces and private entrepreneurship must continue to be given top policy priority. The government should also perform a strategic and coordinating role in the productive sphere.

A developing country is sometimes referred to as a third world, less developed or underdeveloped country. A developing country is generally characterized by less developed industrial and technological infrastructure and a low Human Development Index (HDI) relative to the developed countries. Since the late 1990s, some developing countries have demonstrated higher growth rates than many of the so-called developed countries. Although there is no universally accepted definition of a developing country from that of a developed country, there is sometimes criticism regarding which countries ideally fit into these two categories. There are, however, general reference points such as a country's GDP per capita, level of industrialization and standard of living that help to clarify the distinction.

The case of India and China ideally explains the confusion between the two categories of countries. Although they are regarded as developing countries their phenomenal economic growth over the last several years has advanced the argument for them to be categorized as developed countries. Many countries dislike the terms developing and underdeveloped countries because they connote inferiority to the more developed countries, and in many cases, there is no clear line of demarcation between the two groups since some developing countries have been exhibiting comparable characteristics.

Development Assistance for Small Developing States

There has always been a disparity between developed and developing countries because developed countries possess a greater share of the world's wealth and have been able to develop the infrastructure and industrial systems to promote manufacturing enterprises through which they can dominate trade. To address this disparity between the wealthy developed countries and the seemingly economic stagnation that typifies many developing countries, global institutions such as the WTO and the United Nations have proposed several initiatives. Two notable propositions were "Aid for Trade" and South-South Cooperation designed to help poorer developing countries expand their production and exports of goods and services.

Aid for Trade was conceived in 2005 as a major outcome of the WTO Doha negotiation process. The Doha Round as it was called, was the latest round of trade negotiations among the WTO members aimed at providing well-needed reforms to the international trading system through the introduction of lower trade barriers negotiated through rounds of revised trade talks. The then Trade Ministers lobbied bilateral and multilateral donors to increase the resources for "Aid for Trade," through grants and concessional loans for trade-related assistance to least developed countries to facilitate trade among marginalised countries.

Access to markets in developing countries was a vital component in enabling them to be placed on a path of sustained growth in trade. Through Aid for Trade, a small donor window was created and was designed to help the poorer developing countries particularly LDCs comply with the new WTO rules arising from the Uruguay Round. It was intended that "Aid for Trade" should help developing countries, particularly LDCs, to build the supply-side capacity and trade-related infrastructure (trade facilitation, transport, ports and quality infrastructure) to participate effectively in international trade. It was anticipated that this initiative would enable developing countries to expand the quantity and quality of goods and services they can supply to world markets at competitive prices.

In the context of the Doha Round, at the outset of the Aid for Trade discussions, developing countries sought assurances that increased Aid for Trade would be provided in addition to existing resource commitments to assist them in implementing and benefiting from WTO Agreements and more broadly to expand their trade." Although the Aid for Trade programme had delivered some benefits, some countries have done well while others have not been impacted. For many least-developed countries (LDCs), Aid for Trade has failed to make the contribution that it should by improving economic growth and reducing poverty. This programme has not necessarily provided the assistance needed for developing countries to establish the infrastructure and build human and institutional capacity to enable them to enter the global market.

The second pillar of initiatives to promote assistance for developing countries was through cooperation among developing countries in the southern hemisphere, generally referred to as South-South Cooperation. At a UN conference on technical cooperation among developing countries in September of 1978 in Buenos Aires, Argentina, the conference adopted a plan of action to promote cooperation among developing countries. This initiative became known as South-South cooperation and was designed to facilitate the exchange of information and experience, technology investment and to build capacity among developing countries in the hemisphere. Such cooperation is generally done through governments, academic institutions and civil society organizations. "Developing countries and civil society have repeatedly criticized the way aid is often used as a neo-colonial tool by developed countries, imposing policy conditionalities on developing states and tying aid to commercial, political and military interests of donors."

South-South Cooperation provides an opportunity to forge partnerships that will lead to industrial development, economic growth and poverty reduction. Countries of the south generally lack the resources and technical skills that developed countries in the north possess.

South-South Cooperation transcends foreign aid since its primary focus is on sustainable development. Global challenges like energy, food crisis, climate change issues and poverty have galvanised countries in the hemisphere into strong regional entities with mechanisms for cooperation. South-South Cooperation has come at a time when many development agencies have established a set of human development indices and criteria that have categorized many developing countries in the south as middle-income countries. There is enough evidence to show that such categorization has had its negative drawbacks, for example, many countries have been disqualified from development assistance.

In the absence of sufficient domestic wealth creation and natural resources, governments are generally unable to collect sufficient revenues to meet their financial obligations and to finance public investment and basic services. Having been disqualified from low-income countries, grant financing is limited, hence, governments' other option is to borrow from development institutions at concessionary terms or commercial lenders at less favourable terms. Debt servicing and interest payments then become a major burden on the already fragile public revenues.

Over many years, several regional agreements targeting critical functional areas have been developed from which developing economies have benefited. Among them are:

(i) Economic Integration; The Bolivarian Alternatives for the Americas (ALBA).
(ii) Venezuela Petro Caribe.
(iii) Education exchange; Brazil's South-South cooperation in education.
(iv) Sectoral cooperation; Cuba's support for agriculture.
(v) Technical Assistance; India's technical and economic cooperation.

Many of these programmes are ongoing and they have had a tremendous impact on small developing states of the south. For many years, small island states have dreamt tantalizingly of development achievements that would put them on the path to economic success based on their geographic and cultural peculiarities that would redound to their advantage. These states with relatively small populations held out hope that their excellent tropical climate, proximity to lucrative markets, relatively large diaspora communities and the promise of external institutions and countries offering to assist in their development endeavours were expected to grow rapidly by the end of the 20th century. Very few have succeeded, while most have stagnated. Some have fallen to the scourge of poverty and indebtedness.

Evidently, most small states have their peculiar and distinct development challenges predicated on a plethora of structural problems within the development process, particularly in a Caribbean context where there is a general lack of entrepreneurial culture and world-class firms are few. As it stands, many firms are not equipped with the requisite plant, machinery and other technical requirements to transit to world-class status to face global competition. Hence, they have no other choice than to focus on the domestic market.

Growth in financial services has deteriorated or stagnated in countries where it represents a sizeable portion of GDP, due to greater scrutiny of their banking practices by the OECD countries. There was also a resurgence of the tourism industry from the fallout of the September 11 terrorist attacks in the United States. With the onset of the Coronavirus (COVID-19) pandemic, it is envisaged that it will require several years before the tourism industry fully recovers from the ravages of the virus to the pre pandemic level. The transition to niche manufacturing and high-end services offers much hope for small states. It will be difficult for economies without the necessary skillset, infrastructure and resources to navigate their way through competitive global markets in turbulent times without the necessary policy instruments and the political will to forge ahead. However, there are lessons to learn from other small island states that have managed this transition successfully.

The major opportunity for the future of these island states lies in the revitalization of the manufacturing sector with an emphasis on niche manufacturing as well as the transition to the services industry which can provide opportunities for high-paying jobs. While there is a push towards diversification, given the size of these economies, the goal should be towards adding value to indigenous primary products by pursuing robust innovative strategies and exploiting promising new and exciting clusters. Increased activities in manufacturing can lead to the development of other key sectors. Shifting economic production towards manufacturing may also hold higher promise for long-term growth.

Every country needs development programmes that can support its citizenry over the long term as a means of reducing unemployment and avoiding the scourge of poverty. This is the essence of sustainable development. It encompasses the importance of industrialization with its supporting infrastructure geared towards the achievement of stated goals. No country or region has ever achieved a decent standard of living for its citizens without a robust industrial sector.

Indeed, the merits of industrialization are enshrined in the Millennium Development Goals, which seek among other things, to build resilient infrastructure, promote inclusive and sustainable industrialization and foster innovation. Sustainable development encompasses economies of scale in national output, provision of higher-skilled manufacturing jobs, the expansion of consumption, increased standard of living and setting economies on a sustainable growth path. The expansion of the manufacturing sector is essential for employment creation, as it absorbs surplus labour from agriculture and other traditional sectors.

Chapter 2

The Conceptual Notion of "Small States"

The traditional assumption used in most open economies is that a country is so small that it is a price taker in world markets for its imports and exports and it cannot influence its terms of trade. The main drawback of this approach is that this definition, although theoretically quite useful, classifies several large countries as small, for example, Australia and Canada. Alternatively, one may categorize "Smallness" based on some notion of the minimum efficient scale of output, since most small states are unlikely to have a large enough domestic market to allow them to exploit the economies of scale. "Small" in the context of small island developing states can also be characterized by three indicators viz (1) the ability of a country to influence its terms of trade; (2) sub-optimality; and (3) size measures, using thresholds.

Among the indicators that are sometimes used to define small states include factors ranging from aggregate economic activity to geographic area. There is no commonly accepted definition or approach to identifying the minimum efficient scale. As a result, some authors use arbitrarily chosen thresholds for various social and/or economic indicators such as population and gross domestic product to demarcate small from larger states. Other writers consider a small island to have an upper limit of 10,000 km² and 500,000 residents. They contend that land exists in several political and administrative contexts. Some small island states are politically independent microstates, while others belong to wide metropolitan countries such as Hong Kong to China.

The Commonwealth definition of "small states" is based on demographic size comprised of countries with a population of 1.5 million people or fewer. However, the reference to absolute size is considered on the grounds of demographic, geographic, or economic factors (including different combinations, single or multi-island states). Irrespective of size, small states share the same characteristics; for example, they pursue the same foreign policy, the same objectives of security, prosperity and wellbeing of their citizens. Small states in the wider Caribbean, Pacific, Indian Ocean, Asia, and Africa, represent an integral part of the international order operating in the same broad political and economic environment as all other states. About two-thirds of United Nations members fall into this category. The following factors are characteristics of small island developing states:

1. **Small domestic market size.** As a result of the small size of the domestic market and by extension, domestic demand, the unit cost of production in most small states will be relatively high, as demand restricts the use of a minimum efficient scale or plant. For the most part, there is insufficient critical mass for the uptake of the production output. A country's small size will likely limit the amount of domestic competition since only a small number of firms might be operating in the market, and they may not find it feasible to engage in a particular area of production.
2. **Limited Resource Endowment.** Besides the relatively small population size, most small states are not normally endowed with abundant natural resources and even if they have natural resource endowments, they may not have the capital necessary to exploit them. This explains the reason why many of their indigenous products are sold in their primary state with little or no value-added output.

3. **Small-scale Output and Exports**. Because of their limited resource base, most small states tend to specialise in the production and export of a small range of goods and services, and they are hardly able to satisfy regional and international demand. Without the necessary resources, there is no impetus to enter the global value chain which will require them to expand output and increase export trade. By and large, most small firms are satisfied with targeting domestic markets and at best regional markets.
4. **Open Economies.** Small states are known to depend on trade with other countries to provide goods for local consumption and a large proportion of the inputs used in their domestic production as a result of their narrow resource base. Although a country may have high tariffs, conventional indicators of openness such as exports and imports as a proportion of GDP are usually very high.
5. **Economical Vulnerability.** Generally, small states are more subject to changes in external conditions, and they are usually quite susceptible to natural disasters and other events. The effects of devastating hurricanes on several small Caribbean states, for example, is a case in point. The cost of damage to these islands is significantly higher than the global average in terms of the value of damage per GDP. According to the IMF, the Caribbean is one of the most vulnerable regions in the world to natural disasters. Storms cost on average 1.6 percent of GDP in damages every year, but that figure could be 1.6 to 3.6 times larger due to underreporting of disaster and damage.
6. **Social Homogeneity.** One of the key advantages of a small state is that it usually possess a greater degree of social homogeneity and cohesion that encourages the formation of social capital that indirectly contributes to economic growth. Notwithstanding the fact that there is a greater sense of homogeneity among small developing states, many established cultural norms are gradually changing. There is a growing diversity even among homogeneous groups, which are becoming more inclusive, and

they are embracing members of other cultures. This auger well for social cohesion a cooperation which can redound to social and economic development.

Generally, most of the characteristics that are common to small states could become obstacles to their growth. They are vulnerable to external shocks given the openness of their economies, and their dependence on a limited number of economic activities resulting in greater volatility of their GDP compared to large states. Many developing states rely on specialization in one or two crops to promote international competitiveness. Moreover, their export trade is geared towards regional markets.

Their physical size, small economies and limited market size pose specific challenges that are manifested in higher transportation costs for imports and exports, higher utility costs as well as the exorbitant cost of doing business. On the other hand, size does not necessarily determine the destiny of small states. There are classic examples of small states that have attained economic success. For example, Tuvalu, another Pacific small island state of a similar size and population to Nauru has demonstrated it has the most successful government in the Pacific.

As small developing states contemplate and plan their industrial development strategy, they must be mindful of the vulnerabilities that typify them. There is overwhelming evidence of the impact of environmental and natural events on food production, human settlement, critical infrastructure, human health and safety and biodiversity. Regional economic integration provides a common platform through which member states can better promote their collective interests with the rest of the international community and in multilateral arrangements, often advocating for recognition of their specific challenges.

Country	Geo-Status	Population 2020	GNI per Capita (2020)	Trade as a % of GDP 2020
AFRICA				
Cape Verde	SIDS	555,988	3527	50.65
Comoros	SIDS	869,595	3,330	12.8*[1]
Mauritius	SIDS	1,265,740	22,390	30.0*
Sao Tome and Principe	SIDS	219,161	22,390	-
Seychelles	SIDS	98,462	24,310	77.2
ASIA AND THE PACIFIC				
Bahrain	SIDS	1,701,583	44,330	76.8*
Fiji	SIDS	896.444	10,910	27.6
French Polynesia	SIDS	280,904	-	-
Guam	SIDS	168,783	-	21.8*
Kiribati	SIDS	119,446	4,250	-
Maldives	SIDS	540,542	12,840	69.0*
Marshall Islands	SIDS	59,194	5,130	-
The Federated States of Micronesia	SIDS	115,021	4,100	23.6*
Nauru	SIDS	10,831	20,770	20.0*
New Caledonia	SIDS	271,960	-	-
Palau	SIDS	18,092	19,530	-
Papua New Guinea	SIDS	8,947,027	4,240	-
Samoa	SIDS	198,410	6,480	31.7
The Solomon Islands	SIDS	571,329	2,680	27.7
Timor-Leste	SIDS	1,318,442	4,490	28.2*
Tonga	SIDS	105,691	6,980	21.1*
Tuvalu	SIDS	11,791	6,430	-
Vanuatu	SIDS	307,150	2,880	23.3
THE CARIBBEAN				
Antigua and Barbuda	SIDS	97,928	18,610	42.3
Aruba	SIDS	106,766	-	-
The Bahamas	SIDS	393,248	31,200	39.20*
Barbados	SIDS	287,371	13,010	42.0*
Belize	SIDS	397,621	5,880	55.0

---

[1] *Trade as a percentage of GDP 2019 figures

Country	Geo-Status	Population 2020	GNI per Capita (2020)	Trade as a % of GDP 2020
Cuba	SIDS	11,326,616	-	14.5*
Cayman Islands	SIDS	65,720	58,100	-
Dominica	SIDS	71,991	10,740	38.3
Dominican Republic	SIDS	10,847,904	17,060	18.3
Grenada	SIDS	112,519	14,370	23.1
Haiti	SIDS	11,123,183	2,930	8.2
Jamaica	SIDS	2,961,161	8,850	39.7
St Kitts & Nevis	SIDS	53,192	24,190	-
St Lucia	SIDS	183,629	12,200	-
St Vincent & the Grenadines	SIDS	110,947	12,810	-
Suriname	SIDS	586,634	15,000	64.3
Trinidad and Tobago	SIDS	1,399,491	24,800	-
Guyana	SIDS	786,559	15,370	-

Economic Indicators for Small States

African Small Island Developing States

The African Small Island Developing States is comprised of Cabo Verde, Comoros, Guinea-Bissau, Mauritius, Sao Tome and Principe and Seychelles. There is no sub-grouping umbrella organization that promotes economic cooperation among the six states. Each state continues to be the guardian of its economic affairs. The lack of a regional body to galvanize regional cooperation in several critical areas including trade, energy, single market initiatives and industry seriously constrains these small states. It also deprives them of the opportunity to speak with one voice on the world stage, dissimilar to what obtains in the Organization of East Caribbean States (OECS) in the Caribbean Community among its eight member states. In some states, there is still resistance to open-market policies, which invariably has, to a large extent stymied the growth of these small island states as far as trade and investment are concerned. The private sector is generally affected by institutional weaknesses as well as a lack of efficient legal framework necessary to facilitate development initiatives.

The African Small Island States have not exploited their natural resource endowments through value-added activities. For example, the phosphate from Nauru, and mineral oil, gold, and copper from Papua New Guinea have been primarily exported in their natural state and have not been used locally in value-added activities. Developing high-value niche products for export can create tremendous opportunities for these small states. Small and medium-sized enterprises can be developed using the resources from the agriculture sector to support the tourism industry that is so critical to the survival of these states and for the improvement of the standard of living for the rural poor and marginalised.

For many years, African island states have experienced serious economic challenges which explain their economic situation. Although they produce a wide variety of crops, notably cashew, coconut, cloves, cocoa, coffee and palm oil, almost all these products are sold in their primary form with little value-added. In recent times, some attempts have been made to diversify the economy from agriculture, fishing and tourism and scale up investment in manufacturing. Mauritius, for instance, has broadened its manufacturing base by investing in the production of diamond, jewellery and watches.

In Seychelles, there are small factories that make cigarettes, paints, sheet metals and furniture. Guinea-Bissau which does some processing of sugar, rice and peanuts has not added value to its huge cashew production because the country has not invested in a cashew processing plant and related infrastructure. Hence, it cannot maximise revenue from its cashew production. Comoros which is the world's second largest producer of vanilla, exports one-third of it to France. It is also the world's leading producer of ylang-ylang, a perfume oil which is also exported. Cloves and copra are also part of its export products. These products are excellent primary raw materials on which high value niche products can be produced.

The challenges of smallness and remoteness have combined to limit Pacific economies of opportunities for exporting and participating in global value chains in traditional and new markets for high value and volume commodities. The economic and social transformation has eluded some small African island states because of several weaknesses that generally characterize small states including limited skilled workers, poor infrastructural base, limited technological investment and a struggling agricultural sector.

Targeted agricultural production can stimulate a striving agribusiness subsector from which strong niche market possibilities may emanate to supply markets in which they are unable to compete internationally. Most agribusinesses are small-scale, family-owned enterprises mainly in agro-industrial production such as chilli sauce, banana chips and fresh products including taro and Tahitian lime for the export market.

Traditionally, agriculture value chains that link subsistence farmers with agribusinesses to the global marketplace have been very weak. Small states should redirect primary commodity exports to new niche markets for consumers seeking unique products and new experiences. Accessing these markets and their requirements will necessitate changes in production, processing and marketing. The small Pacific States are renowned for products such as coffee, chocolate, chilli, vanilla, cosmetics and signature clothing which are generally produced at low volumes. Product quality should be given priority to facilitate increased access to international trade.

Asia, Pacific Island Countries (PICs)

The Asian, Pacific Island States consist of 21 small states. The Pacific Islands Forum which was initially known as the South Pacific Forum was established in 1971 as an inter-governmental organization that aims to enhance cooperation among the independent countries of the Pacific Ocean. The Pacific Islands Forum represents Heads of Government of all 16 independent and self-governing Pacific Island countries including Australia and New Zealand. The Forum was established to provide a platform for heads of government to discuss common issues and challenges facing the independent and self-governing states of the South Pacific.

Within the framework of the Forum, the South Pacific Bureau for Economic Co-operation was established in April 1973 to deal with economic matters and to facilitate cooperation among members on matters of trade, tourism, transportation and economic development. Several of the Pacific Island States have experienced moderate levels of development. Fiji with its abundance of mineral, forest and fish resources and Bahrain with its high level of global trade in aluminium, food & beverage, chemicals & plastics have been financial and business hubs in the Middle East and North Africa.

Like many of the other Pacific states, economic development is hindered by several factors including the isolation of these states from foreign markets, the limited size of their domestic markets, the lack of natural resources, periodic devastation from natural disasters and inadequate infrastructure. Many of these states have thriving agricultural sectors that produce a diversity of agricultural products including sugar production, copra, citrus fruit, coconut oil, ginger and citrus fruit.

Some island states possess natural resources including Papua New Guinea with mineral deposits of copper, gold and oil. Timor-Leste produces natural gas which it pipes to Australia because of a lack of processing facilities and Nauru for its phosphate mining. There are also small industries including garment, fruit processing and handicraft that add to their economic diversification. Because many of these economies are struggling, investment is critically needed to build the economy as well as to strengthen institutional capacity, human capital and physical infrastructure. Electricity, telecommunications, roads and other infrastructure that are critical to enabling private sector-led growth remained underdeveloped in many cases.

Pacific Island Countries' domestic markets are generally small and do not allow them to exploit economies of scale. Their comparatively small size and remoteness of major economic centres engender high cost of production which affects the price competitiveness of their exports. Unlike other small developing states which have developed trade integration among members of regional blocs as a means of lowering production and trading costs and have thereby enlarged their markets, PIC has not moved in that direction. The story of how relatively small developing countries like Hong Kong, Mauritius, and Singapore have managed to develop a manufacturing industry that was pivotal to their industrialization, which enabled them to become relatively wealthy countries has not resonated with the governments of these small states regarding policy and strategy.

The Smaller Island States (SIS)

The grouping of Smaller Island States of the Pacific Island Forum, which is synonymous with the OECS within the CARICOM Community is a grouping of eight states which was established in 2005 to facilitate special assistance to those Forum countries that were experiencing limited development capacity within the ambit of a fragile and vulnerable economic environment. The membership now comprises the Cook Islands, Federated States of Micronesia, Kiribati, Marshall Islands, Nauru, Niue, Palau and Tuvalu.

The lack of private sector involvement in matters of development is one of the major factors that affects the growth of the manufacturing sector in the Smaller Island States (SIS). No serious attempt has been made to broaden the productive base to encompass niche manufacturing. Public sector involvement can bridge the knowledge gap and increase the availability of capital which is paramount to the development of niche markets for products and services. The current traditional approach to regulating and protecting markets is a characteristic of inward-looking economies. There is an obvious role for the government to provide the private sector with logistical support and practical market information to develop market intelligence on regional and international markets that can be targeted for external trade. The transportation of raw materials is inhibited because of the lack of efficient airports and seaports.

On the other hand, their geographic peculiarity has limited their economic potential. Lack of telecommunication infrastructure and interconnectivity allied with inefficient transportation systems have also led to sluggish development growth of their natural resource-based primary industries. These drawbacks can be addressed by improving the outdated infrastructure, increasing regional coordination and cooperation among SIS states, and developing striving niche markets. Smallness could translate to unique products that are very limited in supply, while remoteness underscores the unspoiled and exotic nature of raw materials being used. These factors allow Pacific niche products to fetch premium prices in the global market, effectively overcoming the hurdle posed by high production costs.

It must be noted that there are other small states within the European Union, the Commonwealth of Independent States (CIS) and in Europe that are not members of the EU to which reference has not been made.

Structural Development Challenges for Small States

It can be argued that SIDS cannot adequately maximise the benefits provided by a liberalised trade regime until they establish the framework for the development of enterprises to some level of world-class status. Most of these countries have experienced moderate economic growth over the last couple of decades. This level of economic performance was, to a large extent, the result of the shortfall in the agricultural sector, particularly for products such as banana, sugar, and rum, on which the countries depended so heavily.

In recent years, the services sector has overtaken agriculture as the main economic driver of the economy of several small states. In some cases, this has resulted in adverse fiscal implications, fostering indebtedness and posing an obstacle to growth. If adequately supported, the increase in services can have a knock-on effect in boosting private sector growth, particularly in the tourism sector, which will result in economic growth and job creation. There are many development challenges that small developing states face because they lack the resources to propel them on a developmental trajectory.

Small developing countries must overcome many of the challenges that have placed a stranglehold on their development prospects. Some of these challenges are monumental in scope and include limited economic diversification, remoteness from major markets, high transportation cost, limited human and institutional capacity, susceptibility to natural disasters and serious environmental challenges resulting from climate change. After many years of independence, many small states continue to experience severe development challenges beyond their control. Many of these challenges are associated with access to finance for development purposes, access to international markets, equitable participation in the global trading system, declining aid due to the reclassification of their economic status and lack of the ability to attract investment, particularly in infrastructure. Small states generally lack proper infrastructure and the enabling environment for business development. In the absence of these conditions, governments could become an enabler to incentivise investors by establishing the relevant infrastructure that will encourage the development of viable enterprises. Such infrastructure should be designed to attract domestic private firms and/or foreign firms willing to invest in the targeted industries. The idea of establishing traditional industrial parks is debatable in the context of long-term return on investment. However, the issue must be guided by policies that should be informed by market trends.

Continuing trade imbalances and high unemployment in the economies of many of these small states have demonstrated the lack of a domestic industrial base that is necessary for long-term economic stability and sustainable growth.

The competitiveness of most small developing states in the global economy continues to rely on low-cost labour, low-value exports, as well as high-value processed goods and raw materials from external sources. Sustained economic development will require moving beyond elementary comparative advantage and achieving a competitive advantage in export trade. Achieving a competitive advantage in the production of higher value products will create wealth, based on value-added rather than low cost.

It must be recognised that the development of a vibrant manufacturing sector will play a significant role in the ensuing years in the economic and social development of small and marginalised states. The conceptualization of initiatives that will energize this sector should be done against the backdrop of the challenges and difficulties that are being experienced in traditional industries like sugar, rum, and the banana trade vis a vis the uncertainty surrounding their future. Manufacturing export is well known to provide a major source of foreign exchange that enables small states to purchase food and raw materials from countries better endowed with natural resources.

Export trade also provides a means of affording luxury products and technologies originating from more developed countries. These factors undoubtedly play an important part in motivating less developed countries to expand their manufacturing capabilities. Generally, enclave-type manufacturing industries that are designed specifically for the export market should play a critical role in influencing a country's industrial strategies. Such strategies are important in defining, restructuring and transforming the economic activities of some small states from a mono-crop system to multi-sectoral development activities. While some new products are being introduced in specific sectors, others are declining rapidly or are being phased out of production altogether.

Some traditional manufacturing enterprises may survive into the twenty-first century. However, most will disappear as product demand changes and the lack of innovation renders them less competitive. It is incumbent on small states to explore all possible industrial options including the development of non-traditional products and services for lucrative niche markets. Hence, specific areas in which countries are unable to compete at present due to a lack of resources and technological expertise can still be competitive in specific niches without competing directly with large and well-established competitors.

During the decade of the 1990s, there had been a new thrust in some small developing states to diversify into the provision of services and thereby develop a thriving services sector. Although the services sector is fast-growing and offers tremendous scope for economic growth, there has always been a deficiency concerning linkages to other sectors. The agro-processing industry has not fully developed into a strong and viable industry that can utilise the copious supply of tropical fruits and vegetables that are available, primarily to address the demand in the hotel industry. Furthermore, the emphasis on manufacturing has been highly traditional and seems to duplicate many of the activities that are conducive to other countries, thereby minimizing the scope for new product development.

Small states tend to be disproportionately impacted by global economic crises. They often have a slow economic recovery, struggle to meet development goals and many have high debt burdens, which pose a severe threat to their economic growth. Their almost total dependence on strategic imports, notably food and fuel is also particularly concerning. Furthermore, small states are also prone to weather-related disasters, which are anticipated to increase in frequency and intensity with climate change. In some circumstances, a single extreme weather event in a small state can result in damages that exceed its GDP in any given year, thereby reversing development by many years. For instance, in 2017, the vulnerability of small states to external shocks was brought sharply into focus during the hurricane season in the Caribbean which devastated the islands of Antigua and Barbuda, Dominica and Puerto Rico, among others. In most of these countries, the key sectors of the economy such as fisheries, tourism and agriculture that are sensitive to adverse weather conditions were also decimated.

Chapter 3

A Regional Manufacturing Sector in Perspective

Generally, the word "industry" connotes a set of production units engaged primarily in the same or similar kinds of productive activity. Within this context, it is not uncommon to refer to "industry" as comprising a range of activities in the areas of agriculture, tourism, services, and other value-adding activities. The industrial sector is a broad term as compared to the manufacturing sector which is a subset of industry. An industry employs labour and capital and is a distinct branch of trade. Manufacturing is the production of goods using labour, machinery, tools and chemical processes. It involves the transformation of raw materials into finished products.

Historically, traditional manufacturing involved hand-manufactured products, which use basic tools and equipment. This form of manufacturing is still being done today, and in some cases, is aided by the infusion of technology. Hand manufacturing is associated with art and craft, textile production, carpentry, leatherwork, and metalwork, among others.

Over the last couple of decades, the manufacturing sector in most small states has undergone a profound transformation regarding structural development changes, technology application and sectoral linkages and boundaries. As a result of these transformative actions, manufacturing value-added has increased steadily in both industrialized and developing countries since the 1990s. In recent years, premature de-industrialization has also been increasingly noticeable in developing countries, where the manufacturing sector has shown a decreasing share of gross domestic product (GDP). To arrest such a decline would require the intervention of macroeconomic policies and strategies, many of which will be discussed later.

The structural changes that the manufacturing sector has experienced over many decades were due to widespread competition and a trend towards labour-saving technological changes. Small firms can benefit when services are being subcontracted to them, or self-employed persons from larger companies. The nature of employment in the formal manufacturing sector is becoming more heterogeneous than before, as complementary activities and some aspects of the production process are being subcontracted. Subcontracting can lead to a decline in the size of manufacturing firms.

Small developing states have always had a relatively high reliance on the primary production sector, which increases their vulnerability to external shocks because commodity prices tend to be volatile. To build more resilient economies, it is critical to develop a strong, sustainable, and vibrant manufacturing sector. This will require technological intervention with the relevant skillset to boost productivity. This approach will stimulate growth in science and engineering-related disciplines, thereby shifting employment towards highly skilled and better paid job categories.

Caribbean economies for example, continue to grapple with a wide range of challenges that affect economic growth. Most of these island states are vulnerable to natural disasters of one kind or another. With few economic activities from which to create wealth, borrowing becomes inevitable and high debt rates give the islands little fiscal space. Heavy reliance on the domestic market for trade makes it difficult for firms to derive economies of scale and scope in production and distribution. Most of their businesses are focused on local and small niche markets. On the other hand, a small number of SMEs have succeeded in developing export markets and have based their competitiveness on their strong domestic markets.

Manufacturing is a catalyst for economic growth through forward and backward linkages. Backward or (upstream) linkages are created when a sector utilizes the products from other sectors as inputs, for example, the growth of the cotton industry may encourage the growth of the textile industry. The auto industry has a direct backward linkage to the steel industry and an indirect backward linkage to the coal and iron industries since coal and iron are inputs to steel production.

On the other hand, forward (downstream) linkages are created when the growth of an industry leads to the growth of the industries that use its output as input in their production processes, or when the output of an industry helps propel another industry. In so doing, the growth of one sector can fuel the development of other related sectors.

Since 2005, the contribution of manufacturing to total value-added in the Caribbean has contracted significantly. The average decline in contribution has been around 1.2 percent in the 2000s. A decline in the region's competitiveness is one of the main explanations behind this fall in the contribution of the manufacturing industry to total value-added. As manufacturing in the Caribbean has contracted, so have employment opportunities.

Historically, most small developing states have not been reputed for high technology manufacturing activities such as the manufacture of automobiles except in the 1980s, when a few developing states assembled cars, domestic appliances, and computers. Regional manufacturing has been restricted to six industrial areas viz. food and beverage, textile, garment, footwear, printing and packaging, wood and related products including furniture, chemicals, cosmetics and pharmaceuticals and assembly type related industries.

The Caribbean food industry can be broadly categorized as consisting of two groupings of enterprises: large and medium-sized enterprises and small and micro enterprises. Generally, large and medium-sized enterprises are involved in the production of sugar, soft drinks, beer, rum and a diversity of processed products from the local raw material base, as well as from imported raw and semi-processed materials. Small and micro-enterprises on the other hand, produce a diversity of products of variable quality, primarily from the local raw material.

Moreover, it is also important to mention that a high proportion of the region's small and medium enterprises are mainly family-owned businesses operating on a day-to-day basis without any clearly defined strategy for production and sustainability. Within the OECS for instance, there are few limited liability companies, and most business activities are undertaken by small and medium-sized family-owned enterprises. Over 75% of the companies operating in these small states are micro-enterprises with five or fewer employees. The sole proprietorship is the dominant type of business operation.

There is a high level of informality surrounding the structure and functioning of these enterprises. They are generally not overly concerned about products or process improvement to penetrate new markets. Intra-firm relationships are rare among these small and medium-sized firms except in sub-sectors where some form of cluster arrangement has been introduced. Well established cooperation with suppliers is not the norm, but in some situations, depending on the product being produced and the level of sophistication of the clients, the establishment of a proprietor supplier relationship is paramount. The scope for backward and forward integration of industries is therefore extremely limited.

In delineating the profile of the manufacturing sector in the Caribbean region, the smaller territories of the OECS, the Bahamas, Barbados, Belize and Guyana have no real manufacturing industry of significance today. The manufacturing sector has declined over the past several decades, with the disappearance of assembly type manufacturing, garment production and similar activities. Manufacturing in the region is primarily dominated by food and beverages.

The Jamaican manufacturing industry has also declined significantly over the last couple of decades. Although it is still a major contributor to the Jamaican economy, its share of total production has been on the decline. In some countries, while the manufacturing sector appears to have grown in nominal terms, it has not grown in real terms. On the other hand, Trinidad and Tobago has the most vibrant manufacturing sector in the region and has become a major extra-regional exporter primarily due to the oil boom at the time. Food and beverage accounted for half of Trinidad and Tobago's manufacturing, with liquid natural gas and petrochemicals not included.

Between 2005 and 2008 the contribution of manufacturing to regional GDP has been constant ranging between11.6% and 12.3%. However, there was a sharp decline of 1.4% in 2009, but increased again for the next three years, while in 2013, it declined to 10.8%. The decline continued up until 2015 at an average of 0.5 %.

Except for Belize, St. Lucia, and Trinidad & Tobago, all the other CARICOM countries have experienced a decline in the contribution of manufacturing to GDP between 2000 and 2007. In the wider context of industrial development, the services and tourism sectors have outgrown manufacturing. Some countries are beginning to diversify away from manufacturing to emerging sectors like services and information and telecommunication. The prospect of achieving rapid growth in these newer sectors with less volatility over a longer period is promising. Unemployment often reflects an economy's inability to adapt to change and the degree to which an economy is adaptable depends on the nature of its institutions, structural policies, regulations, laws and social and cultural values.

An Analysis of Fiscal Incentives Provided

The increasing mobility of international firms and the gradual elimination of barriers to global capital flows have encouraged competition among governments to attract foreign direct investment, often through tax incentives.

Fiscal incentives involve those policies designed to reduce the overall tax burden of a foreign investor. In a broader context, fiscal incentives include tax holidays and import duty exemptions to investment allowances and accelerated depreciation have been used globally as a strategy to attract multinational companies. It is anticipated that there will be enormous technological spillovers from which countries can benefit. This trend, which began in the 1990s, has been strengthened in succeeding years.

The financial incentives provided to foreign firms typically include relief from income and dividend taxation, reduction or exemption from import duties, property tax concessions and other financial incentives such as exemptions from exchange controls and stamp duty on land purchase, exemptions from import duties on raw materials, intermediate inputs and capital goods. A few countries have chosen a non-targeted approach by lowering the effective corporate tax rate for all firms while providing limited or no incentives. Some countries also provide training grants and research and development subsidies. It would appear that the benefits from these concessions are uncertain, while the cost is exorbitant. The incentives available in most regions are very generous and the conditions for qualifying for them are often not strict.

Selected Incentives Provided by Some Small States

The following is a list of other major incentives offered to manufacturers in various small island developing states.

- Subsidized rent for factory space in well planned and fully serviced industrial parks.
- Waiving of General Consumption Tax (GCT) chargeable on the acquisition of machinery and equipment directly related to the manufacturing process as part of the industry modernisation.
- Waiving of the Customs User Fee on capital equipment and raw material for a specific period of operation of the entity.
- Exemption from Income tax on profits in perpetuity, import duty and licensing fee in perpetuity, repatriation of foreign exchange without any form of recourse on the part of the government.
- Provision of Loan Guarantee which provides financial support to existing and expanding businesses and new or start-up businesses.
- Reduced time frame (in some cases) for depreciating the cost of capital equipment.

- Export Grant Credit Scheme which covers pre-shipping, financial requirements and post shipment credit risks.
- Taxing offshore companies at a rate significantly lower than the normal rate for local companies.

Several governments use investment incentives as a means of attracting investments into their countries. There is an ongoing intense debate regarding the effectiveness of incentives and in particular tax incentives on foreign direct investment. There has been little consensus on the impact of these incentive schemes on investments. Some experts have argued that there is little evidence to suggest that such incentives are effective. In light of this point of view, there have been recommendations proffered, that governments should curtail the use of these incentive schemes. On the other hand, there are those who argue that investment incentives have contributed significantly to the rapid economic development and growth of some countries such as the Republic of Korea, Mauritius, and Singapore.

Notwithstanding the diverse views on this subject, it is a common approach that tax and non tax incentives are just one of the many factors that influence the success of investments. It is also the norm that countries typically pursue growth-related reforms using a combination of approaches including macroeconomic policies, investment climate improvements and industrial policy changes, among which are investment incentives. If such reforms led to growth, it is difficult to attribute such growth solely to incentives.

The most typical tax incentives are tax holidays, special zones, investment tax credits, investment allowances, accelerated depreciation, reduced tax rates, exemptions from various taxes and financing incentives. These incentives are seldom used in isolation and many governments employ a combination in their package of fiscal incentives. Contrary to the view that the main stimulus for investment in industrial development should be linked to an attractive incentive scheme, it must also be noted that an effective incentive scheme should be designed within the framework of a favourable and enabling environment with all the available elements of support to develop and promote the infusion of new ideas and innovation. The importance of the private sector must be recognised and encouraged to achieve growth within the economy. The government must also understand and fulfill its role by creating the environment within which that growth should take place.

It is being reported that 50% of the construction sector in a small island state is dominated by the Chinese of which 50% of the labour force on those construction projects involved Chinese workers. Most of the companies are wholly or partially owned by the Chinese government and they are well resourced. These companies receive concessionary financing from the government including duty-free concession on imported machinery and equipment and the purchase of building inputs free of General Consumption Tax (GCT). They then use the advantage gained through these concessions to compete with local contractors for private sector projects.

Tax Holidays

Tax holidays are temporary exemptions from specific taxes (normally the corporate income tax) and may sometimes include exemptions from administrative obligations (such as the need to file tax returns). Tax holidays may also be partial if they offer reduced requirements rather than a full exemption. Most Caribbean countries, for example, either had or presently use tax holidays to attract investment, particularly FDI, which could lead to knowledge spillovers and economic growth. In its assessment of tax concession, the Asian Development Bank was critical of the Fiji government's policy of awarding tax concessions to large multinational companies investing in Fiji. It stated in its 2005 report that concessions have been abused and have not generated long-term investment. The report also accused foreign entrepreneurs of leaving as soon as their concessions have expired and alleged that the administration of the concessions encouraged corruption and bribery.

Tax holidays are among the most widely used incentives, especially in small developing countries. They provide benefits as soon as a company begins to earn income, while the benefits of a lower corporate tax rate accrue more slowly and over a longer period. Short-term investments are particularly prime beneficiaries of tax holidays, primarily "footloose" industries in which companies can move quickly from one jurisdiction to another. Fiscal experts have generally been highly critical of tax holidays because they tend to erode the tax base of a country, especially where there are widespread evasions of income taxes.

Newly founded companies tend to benefit more than companies that are in operation over a long period since these companies will have time to establish themselves in the market without the pressure of a heavy tax burden. It is generally recommended that incentives should only be used temporarily. Tax holidays, if used, should have an end date after which they are not available to anyone.

Over the past few decades, results of time-series econometric analysis and numerous surveys of international investors have shown that tax incentives are not the most influential factor for multinationals in selected investment locations and jurisdictions. Economic analysis and surveys have confirmed that tax incentives are poor instruments for compensating for negative factors in a country's investment climate.

However, it does not mean that tax incentives have no effect on foreign direct investment. It is difficult for small enterprises to compete in a globally competitive environment in the absence of FDI. Hence, small states must devise attractive offerings that will lure foreign companies to invest. Within this context, many states have introduced various incentive-based measures including the use of tax incentives to promote FDI as one of the major strategies to promote their comparative and competitive advantages. Furthermore, there are numerous policy driven initiatives that are critical to the success of FDI, some of these include stable and sound economic policies, a stable political environment, supporting legislation and institutions and the development of human capital. Tax incentives are useful only in cases where they are carefully used and monitored and when they become an integral part of a package of sound economic policy measures and strategies.

Foreign Direct Investment is a strategy used by many governments to accelerate the pace of development in critical sectors of the economy. FDI can be a catalyst in bridging the digital divide between developing and developed countries. One of the ways of attracting foreign firms to a host country is for governments to offer incentives of varying kinds to foreign firms. Financial incentives often involve providing funds directly to the foreign investor, for example, investment grants and subsidized credits.

A tax incentive can be described as a deduction, exclusion, or exemption from a tax liability offered as an encouragement for businesses to be engaged in specific investment activity for a specified period. Tax incentives can be given in a variety of ways including investment tax credits, investment allowances and accelerated depreciation.

These tax allowances can be provided in three ways: (i) Investment tax credit which is a tax-related incentive that allows individuals or companies to deduct a certain percentage of specific investment costs directly from their tax liability, apart from the usual allowances for depreciation. (ii) An investment expenditure allowance is a tax incentive offered to businesses to encourage capital investment in which they can deduct a specified percentage of the capital cost, including depreciation from their taxable income. (iii) Accelerated Depreciation incentive is an incentive that allows companies to write off assets faster (for tax purposes than for accounting) than using straight-line depreciation. Two main methods employed in accelerated depreciation are (1) Declining balance depreciation and (2) Sum of the years' digits depreciation. These two methods allow a firm to write off a larger proportion of an asset's value in the early years of its life than in the latter years. The justification for accelerated depreciation is that an asset is more valuable and has more earning potential in the early years of its useful life. In later years, the asset becomes less valuable due to its inevitable obsolescence, and deterioration because of wear and tear.

Investment tax credits and investment allowances both involve the deduction of a certain fraction of the investment from taxes and allow companies to reduce taxes paid by a percentage of their investment expenditures. Many jurisdictions including those of the Caribbean have started to incorporate these types of incentives in their industrial policies. Investment allowances and tax credits have been noted for transparency, easy implementation as well as being contingent on new investment. They also have several disadvantages including the risk of eroding the tax base if they reduce taxes on existing investments.

Furthermore, investment allowances tend to direct capital towards short-lived goods, as any replacement will once again qualify for the allowance. Allowances of these types are usually applied to physical investment and therefore provide a bias against financial and human capital development. These incentive schemes are, nevertheless, useful for projects that have low profitability rates and are superior to tax holidays given that they would at least earn some government revenue.

One of the shortcomings of tax holidays is their diminishing effect on investment, partially after the initial investment at the beginning of the tax holiday, as any additional investment will attract increasingly higher rates, making tax holidays particularly attractive for front-loaded, rapidly profitable investments. These incentive schemes are particularly useful for profitable investments, rather than generous depreciation schemes, which would be more attractive for low profitability schemes. It has been recognised that tax holidays have eroded the tax base in many Caribbean countries.

Specialised Arrangements for Business Development

Traditionally, many governments have provided physical infrastructure such as industrial parks and free trade zones as investment strategies to lure foreign direct investment. More recently, technology parks or cyber parks have been developed to encourage and accommodate high technology firms. Whether these spatial arrangements have provided the anticipated benefits is a matter of debate. Export processing zones and industrial estates require subsidization of infrastructure and land to encourage investments. Apart from these programmes, developing states are more likely to base their incentive schemes on tax holidays and other fiscal measures that do not require direct payments of scarce public funds. Moreover, there are no reliable calculations of how costly these incentive programmes are. In the absence of published FDI subsidies, it is difficult to determine the different kinds of incentives that influence investment flows and firms' behaviour. Whereas subsidies may influence the location of firms, tax holidays may affect operational decisions for many years especially when they are running out.

Generally, there are doubts about the cost-benefit of financial incentives, and there appears to be little evidence that suggests these incentives are critical to attracting investment. It can be argued that if the business environment was improved significantly, there may not be any need for policies for FDI promotion. From the policy and administrative side, proactive and effective government intervention streamlined to include less bureaucracy, better public services, a faster judicial system and better contract enforcement would make it easier to do business that will benefit domestic and foreign investors alike. If such improvements in the business environment were to be implemented, then the pressure to offer greater incentives for investment would dissipate or would become easier for governments to shrug off. Providing incentives to attract investment is not necessarily a bad notion. However, financial and other incentives may not necessarily result in increased innovation among enterprises which is necessary to foster competitiveness.

Special Zones

Free Trade Zones (FTZ) are geographically limited areas in which qualified firms can locate and thus benefit from an exemption of varying scope of taxes and/or administrative requirements. They are often located near major ports that offer companies a variety of financial incentives from the local government to encourage investment in these areas. There are strategic arrangements designed to attract direct foreign investment, which will generate foreign exchange, provide employment opportunities and contribute to economic growth. The most common incentives include duty-free imports for vital business inputs and tax exemption on corporate profits. Some areas where qualifying companies operate can be declared 'zones' irrespective of their location. It was envisaged that EPZ would create indirect benefits through spillover effects due to the geographical proximity of similar firms. FTZs were also established to attract Foreign Direct Investments that would generate foreign exchange and serve as sources of employment to bolster the country's economy. FTZ can be grouped into the following three major categories:

- Those that focus on manufacturing often entail the assembly of imported inputs, such as textiles, garments and light manufacturing.
- Those that are IT-oriented attract businesses in high-tech industries and provide relatively higher paying jobs.
- Highly integrated Cyber Parks, or techno parks as they are commonly called, that aim to provide the necessary components for business development and technological innovation.

When FTZs became popular in the 1980s, the focus was primarily on the manufacturing sector such as textiles and garments. Developing countries across the world have experimented with different models of the FTZ as a strategy to attract foreign direct investment. Countries that applied this model competed primarily on cheap labour. The most notable example of the practice of FTZ models is the maquiladoras found throughout Latin America. The impact of this incentive scheme includes reduced funding for social services, limited transfer of skills to locals (most locals employed in these zones in the Caribbean for example, were offered low-skills jobs), and the high probability that the firms are pursuing goals that conflict with the development strategy of the government.

The success of this strategy concerning employment generation is less than impressive, with employees of export processing zones accounting for as low as 2 percent of the total workforce. The idea of establishing FTZs is a strategic one. It is geared towards attracting FDI which would provide a series of economic benefits. Export Processing Zones have been established on several islands of the Caribbean in the 1960s. St. Lucia was one of the earliest countries in the Caribbean to set up EPZs when in 1968 a US manufacturer using coils wire for transformers and fields of electric motors began operations on the island. Similar enclaves were set up in other countries including the

Dominican Republic, St. Vincent and the Grenadines, Jamaica and Trinidad and Tobago.

Export processing zones have traditionally been established to facilitate manufacturing for the export market. They have received tremendous benefits from the state over firms operating under more regulated conditions. Among other things, they have benefitted from duty-free imports of raw materials, less bureaucratic red tape, generous concessions including long term tax holidays and subsidized rent and utilities, even in times when the cost of such services has been skyrocketing.

After several years of operation, most of the countries in which they operated have not benefitted in any significant way from EPZs and were forced to abandon this policy. It is time for governments to examine the usefulness of these instruments in light of the dynamics of global trade. For EPZs to succeed in the current market environment, the entire model needs to be revamped to allow for more flexibility in the rules under which they are governed. For example, restrictions on domestic sales, inflexible regulatory environment and tax holidays should be revised to give way for open competition. Governments now face several options to phase out the FTZs; among them is the establishment of a business-friendly environment, sound fiscal policies, and industrial policy with all the trappings of infrastructure development, incentives to improve trade, technology transfer, innovation and other measures that will augment firms' competitiveness.

Industrial Parks

The term "industrial parks" is often used in connection with a broad range of concepts including, free-trade zones, export processing zones, special economic zones, high tech zones, enterprise zones, etc. However, the use of the term "industrial park" is generally based on the objectives, functions or forms, differences in the economic policy and terminology of various countries. The variation in terms and concepts associated with industrial parks is often a strategy used to differentiate themselves from the competition. These parks can be structured to provide complementary services to other industries. They could accommodate a mix of industries and facilities including ports, warehouses, distribution centres and factories. Some industrial parks could offer tax incentives to encourage business development.

Industrial parks have risen to prominence in the 1960s when an increasing number of countries embarked on the promotion of industrialization and economic restructuring through the industrial parks' initiative. Most industrial parks were established to support the export industrial strategy of the country in question. An industrial park refers to a parcel of land subdivided and developed within an area that is zoned exclusively for industrial use. They offer, as a minimum, a site "package" to the potential industrial user, along with protection of the investment on the site. The amenities and site improvements offered include at least the following: (i) relatively good location, (ii) utilities, (iii) transportation and (iv) services. Not only are they provided with pertinent physical infrastructure such as transportation, power and water supply, but industrial parks should also be provided with pertinent innovation infrastructure such as relevant technical support and training facilities. Provision is made for roads, transportation and public utilities to be used by manufacturers.

In the case of developing countries, industrial parks can maximize resource integration for limited production factors within a certain spatial scope. By attracting labour and capital-intensive domestic and foreign investment in manufacturing and service industries, industrial parks can increase job opportunities, wages and skills of local workers. Furthermore, they can also establish links to global value chains by participating in international competition and making full use of comparative advantages to promote the upgrading of industrial structures. In so doing, the country's position in the international division of labour can improve. Currently, the industrial park economy has become a global trend.

Industrial parks can be found in most small developing states with different variations and configurations based on national objectives and strategies. The establishment of these parks is one of the strategies which will enable intensive and efficient use of limited land. They can be an effective mechanism to promote industrialization and the structural transformation that is associated with them. Provision for the development and growth of industrial parks should only be geared towards homogeneous businesses that fall under specific target industries. These dedicated facilities are designed primarily to overcome high production and transaction costs that are financially burdensome to small firms.

However, there needs to be proper implementation of these facilities to gain the benefits that are intended. By delivering these public goods and the accompanying policy interventions to support investment, industrial parks have been a catalyst in facilitating industrial development as in the case of the newly industrializing economies (NIEs). Most industrial park programmes that are designed arbitrarily without clear focus and objectives often lack effectiveness. For the most part, they are market-driven, in that, they respond to actual investment demand and investor needs. A "build it and they will come" approach only works in those exceptional cases where there is significant pent-up demand for land for industrial purposes, commercial facilities, and where industrial park location matches firms' expectations.

Despite several advantages of industrial parks, there are some limitations that can be overcome with proper planning. One concern is that only those enterprises that offer promising prospects may be targeted for selection for space in the industrial park. In so doing, firms with high growth potential can be overlooked. Bearing in mind that targeted public support should be confined to activities within the economy's existing comparative advantage. This could be used as a defence against the accusation that any sectorally targeted support amounts to "government picking winners."

Cyber Parks

Technology parks sometimes referred to as "Cyber Parks" are generally established within the ambit of an industrial park. They are key infrastructure in supporting the growth of today's global knowledge economy by providing a location in which government, private sector and universities cooperate to develop products and services of a technological nature. These parks create environments that foster collaboration in innovation and technology. They enhance the development, transfer and commercialization of technology from targeted research.

The advent of Cyber Parks in the 1990s provided the opportunity for the development of high technology industries driven by innovation. The full integration of successful cyber parks has given rise to self-sufficient towns that encompass residential neighbourhoods, shopping centres, universities driven by technology, research centres and business incubators. Many successful Cyber Parks exist throughout the developing world including Taiwan, Indonesia, Malaysia and the Philippines. Similar projects have begun in Latin America in the case of Costa Rica and the Dominican Republic. Most Cyber Parks aim to replicate what Stanford graduates Hewlett and Packard achieved in Silicon Valley in the early 1990s.

Cyber Parks rely heavily on the sustained creation of significant intellectual capital through universities, research centres and business incubators, all of which are costly and cannot be developed overnight. Due to the complexity of design and implementation, successful techno parks require patience and unwavering commitment from the government. It can take upwards of 15 years for the model to fully mature to a stage where all its components are pushing towards a common objective which is to enable innovation and increase competitiveness through cooperation.

Cyber Park projects require massive investments in infrastructure. The Taiwanese government invested roughly US $483 million in developing the infrastructure of its 1,000 acres Hsinchu Science-based Industrial Park over a 15-year period, including two major universities and the Industrial Technological Research Institute. Four national research laboratories are also located in the Hsinchu area. The Technology Park Malaysia (TPM), inaugurated in 1996, covers 120 acres (50 hectares) and involves a state investment of US $80 million and is surrounded by five universities.

Developing an industrial park does not necessarily create any value unless the park is additionally equipped with the requisite physical infrastructure and other enabling conditions that attract foreign manufacturing firms. Hence, for many small states like those of the Caribbean, Pacific and Africa and Asia, such an initiative may likely lead to failure. It is sometimes mythical to think that industrial parks automatically generate economic benefits. Moreover, governments that are not adequately sensitized to the rudiments of industrial parks, particularly in meeting the requirement of firms tend to over expend resources on building these "empty shells," which only a limited number of firms may be prepared to use. A large amount of planning, skills and effort are required to make an industrial park effective in fulfilling its objectives.

Chapter 4

Technology Transfer Towards Competitiveness

It can be argued that many SIDS will find it difficult to compete directly in the global market with developed countries that have already established a strong technological base and possess inherent advantages relating to their size and resources. The history of competitiveness in small island states bears evidence of the fact that success is achieved when interventions are made in products and processes that are already endemic in society. To survive in the marketplace, small developing states should develop a manufacturing culture that is responsive to the challenges of the global markets which is driven by innovation in technology, product design and marketing.

There are two options facing small and medium-sized enterprises in their development process (i) use the result of research from their R&D department to develop innovative projects to meet market needs, or (ii) they can choose to purchase the technology through the process of technology transfer. A useful strategy in the absence of research facilities and resources is to align themselves with research institutions such as universities and specialised institutions that focus more on research for commercial applications.

For most small enterprises, new initiatives are required to stimulate growth in the manufacturing sector. It is against this background that this chapter will focus on several initiatives that would enable the industrial sector in most SIDS to produce efficiently by maximizing its resources. Enterprises will therefore have a greater chance of competing successfully in the global market. The two initiatives that will be discussed are the concepts of clustering and incubation for start-up firms. These are known to offer tremendous benefits to small and micro firms if they are properly implemented. These concepts are not new but there have been variations in the results based on the implementation modalities.

This chapter will examine the extent to which technology transfer could be used effectively in the development process. Commercial and marketing innovation will also be promoted as a means of making products more marketable. Niche marketing is being proposed as an option to be pursued for sustainability in traditional markets. Innovative marketing and new product development will be discussed in detail regarding competitiveness and commercial success.

Technology Transfer; A Development Tool

The term technology transfer has no universal definition. However, there have been various appropriate definitions despite their definitive variations. Technology transfer can be conceived as the acquisition, development, and utilization of technological knowledge by a country other than that in which the knowledge originated. It involves a process of transferring knowledge in some form from a person or organization who possesses it (the transferor) to another person or organization who arranges to receive it (the transferee). The term can be more meaningfully applied in a wider context to involve all aspects of the transfer of scientific or technological knowledge from one society to another to satisfy the identifiable needs of customers.

Small developing states can hardly exist in isolation from developed and advanced economies such as Japan and countries in Europe and North America that have the resources and technical know-how to foster research and development in science and technology. Few countries are self-reliant on technology; hence, over the last several decades the global transfer of technology has become an inevitable phenomenon across frontiers and has enabled the production of goods and services for local, regional and international markets.

The transfer of technology is pivotal to the social and economic development of any nation. For example, the economic prosperity of the United States, Japan and Germany is closely linked to the development of new technology driven by innovation. Technology and innovation are considered the most important factors in improving productivity, quality, and promoting competitiveness. Some of the most powerful industrialised countries like the USA, China and Japan have adopted massive technology programmes from which their economies have benefited substantially. However, other countries have found it unjustifiable to commit valuable resources to scientific research without "reinventing the wheel." Their phenomenal record of reengineering or modifying foreign technology to meet the needs of different sectors within the global community is evidenced by their prowess in many aspects of modern technology. Science and technology is increasingly becoming a critical element in the development toolbox of small states.

In the context of the global economy, a nation's economic success rests mainly on the competitiveness of its enterprises. In the prevailing global economy, it is incumbent on governments to focus on devising activities that will foster enterprise development through the creation of linkages with Science & Technology driven institutions. Since there is heightened interest in small manufacturing enterprises (SMEs), there is a need to develop entrepreneurial promotion schemes as well as research and training facilities in manufacturing technologies to support this sector.

In the process of building technological capacity, technology should not be considered an end in itself. On the contrary, it is a tool to be used to achieve economic and social development goals. In the recent past, many governments have made attempts to strengthen the small business sector by initiating several measures including the establishment of development banks or similar institutions to assist in providing funds through soft loans as well as technical assistance to potential entrepreneurs. Bureaus of Standards were developed and strengthened to ensure that products meet quality requirements that will qualify them for the export market. More recently, entrepreneurial development centres or what is commonly known as enterprise/incubation development centres were established to support small businesses in several technical areas until they are well established as economic entities. Although these initiatives are commendable, there is little effort at attempting to link science and technology to innovation and creativity in the productive sector.

To build technological capacity, large pools of scientifically and technologically trained personnel are necessary. While investing in developing such manpower pools, it may be necessary to involve industry in national manpower planning exercises, education and training and R & D policies. It is important for every country to make decisions in response to the new paradigm in trade and competitiveness. Where there is reliance on foreign technology, small states can develop the capacity for gathering information and forecasting major technological changes that affect principal export products.

It can be argued that the lack of technological capability is one of the major drawbacks in achieving a higher level of industrial development in those small states. These states have been pursuing some level of industrial development programmes since the early seventies. More importantly, the impact of such technological shortcomings has been evident in the establishment of enclave "offshore" enterprises and the promotion of import substitution programmes. Having gone through this phase of development with very limited success in terms of the economic linkages created and the transfer of technology through the lack of foreign direct investment, some countries are now beginning to place greater emphasis on an export-based industrialization programme predominantly on indigenous enterprises.

In the absence of some coordinating mechanisms at the planning level, policies and programmes will be devised on their own and in a vacuum. Consequently, scarce resources will continue to be wasted and their contribution to development will not be maximised. Within the public sector, there are several science and technology-related activities that form part of the regular development programmes of many countries, which for the most part reflect immense duplication within departments. As a result of this approach, small states continue to compete for scarce resources to pursue the same programmes, albeit in a piecemeal and uncoordinated manner. In addition, the education sector in many small states has its own science and technology agenda (dictated by the requirements of external examinations) without any direct linkage to or focus on national development needs.

Building Technological Capability

In modern societies, technology is impacting everyone's life in every sphere of endeavours in ways never experienced before. Every country is in the quest of acquiring some form of advanced technology, either in factories, the transportation system, telecommunication, energy and the like. In other words, some countries are committed to building technological capability to advance their development programmes. Technological capability is the entire complex of human skills (entrepreneurial, management and technical) needed to establish and operate industries efficiently over time. Technological capabilities are not developed fortuitously. Although some breakthroughs may occur serendipitously, there must be strategies aimed at developing the institutional infrastructure and capability that would encourage and promote innovation through a process of transformation into marketable outcomes.

Successful technology transfer should be predicated on a widely developed educational system supported by sound scientifically based programmes designed to encourage creativity and innovation. Such programmes should be the catalyst for the development of requisite skills and competence that are necessary for the effective implementation of technological programmes in any given environment. Many developing countries are endowed with latent creative abilities. However, transforming creative ideas into practical and meaningful development projects could be inhibited by serious structural and economic factors.

The educational programmes in every country should be assessed periodically to ensure that they are aligned with the needs of the country in question and with international norms. Scientific disciplines should be less theoretical and should incorporate adequately designed laboratory and practically oriented research programmes to allow academics and technical minds to challenge established principles and practices and to seek alternatives to traditional thinking. Upgrading the educational skillset of the population as well as paying special attention to human resource planning and development designed to address the scientific and technological needs of a country should be a priority.

Research institutions, polytechnics and other technical institutions should be involved in developing the manpower capability of a country by designing and offering specialised tertiary education programmes including those for practising managers. Specialised databases can be the source of technical information relating to new technological developments. Research institutions should be adequately resourced and strengthened to provide R&D services to agencies that use them. The dissemination of information can best be facilitated through extension services that will function with reciprocity in which information can be disseminated to the end-user of technology and through which research institutions can receive feedback from the user of technology on issues identified to be addressed.

Scope and Dimensions of Technology Transfer

Successful technology transfer is most likely to be influenced by several factors apart from a country's resource endowment. The transfer process should take place within the confines of what is appropriate to the needs of the recipient country. It should be concerned with all aspects of the development of a community leading to an integrated approach to economic and social development. The transfer of technology should have a relationship and compatibility with all existing cultural, economic and social elements of the society in which it is concerned. Hence, the following factors should be considered in effectuating the successful transfer of technology:

1. **The geographic reality of a country**: It is pointless to obtain the requisite technology for specific applications when a country lacks the basic resources that its geographic reality does not support, and which would otherwise have impacted positively on its production endeavours. Hence, the geography of any country can influence its technology transfer possibilities. As an initial consideration, a country should therefore understand its geographical strengths relative to the products it wants to produce or the services it wants to offer. In this regard, investment in technology must be carefully considered with respect to this matter.

2. **Established Cultural Norms:** The cultural value system of a country should also be understood, rationalized and integrated into the planning process that would enable the successful transfer of technology. This is especially critical for developing countries where pervasive cultural norms are decisive factors in how people accept and react to new technology. Culture is a highly sensitive issue and people and culture are inseparable. It is important to analyse those facets and practices of social life that are of value to the people within a specific locality or setting, to understand what will motivate them to work as well as to change their behavior in a particular way. For instance, more traditional oriented people will resist new technology to protect their culture, while more futurist cultures will strive for new development.

   Against this background, education and training is important in establishing a scientific and technological culture that will readily respond to new technological challenges. Technology can negatively impact the cultural value system of a society years after its implementation. Any negative effects should be anticipated in the planning process and provisions should be made to avoid them or mitigate them in circumstances where they cannot be avoided.

3. **Government as an Enabler**: Small developing states are often deficient in the infrastructure requirements that will enable

them to transfer new technology and break the vicious cycle of underdevelopment. The government can act as an enabler by creating the appropriate environment to encourage the business sector to invest in technological activities by making the necessary regulatory provisions for ease of transfer of the technology.

4. **The Economic and Social Landscape**: The economic needs of poor states should dictate that resources be expended in the areas of priority such as health care, education and social services rather than to transfer technology for advanced scientific purposes beyond the capability of the country. The converse is true of advanced economies that possess the capacity to further strengthen their economic status, irrespective of the strength of the economy. Some small countries with thriving economies establish limits to which they can transfer technology. For the most part, they have concentrated on research niches where they have developed comparative advantages. For instance, Finland leads the world in basic and applied paper research, Switzerland is outstanding in chemicals and pharmaceuticals, Austria is advanced in steelmaking and New Zealand in dairy products.

5. **Enabling Business Environment**: Without entrepreneurial commitment, the transfer of technology cannot take effect. A business environment that fosters innovation and promotes entrepreneurship can be the catalyst for nurturing a positive entrepreneurial spirit. Entrepreneurship can be used as the vehicle through which technology can be successfully transferred. While entrepreneurs pursue profitable returns on their investments, they must improve their operational efficiency by adopting new technology. These goals are possible if the environment is conducive to business development.

Technology Embodiment Scenario

Technology can be manifested in tangible and intangible forms. The tangible forms of technological embodiment include those physical objects that are seen everyday such as equipment, machinery, physical facilities, through design features. (i) Technology embodied in institutions is manifested in the form of management systems, designs and networks, etc. (ii) Technology embodied in human form accentuates the attributes imbued in human through people's ingenuity, creativity, knowledge, skills and experience, etc. These are manifested through developments that pervade the physical landscape (iii) The embodiment of technology in information manifests itself as document related including facts and figures, publications, blueprints, software that includes the know-how such as processes, techniques, methods, etc.

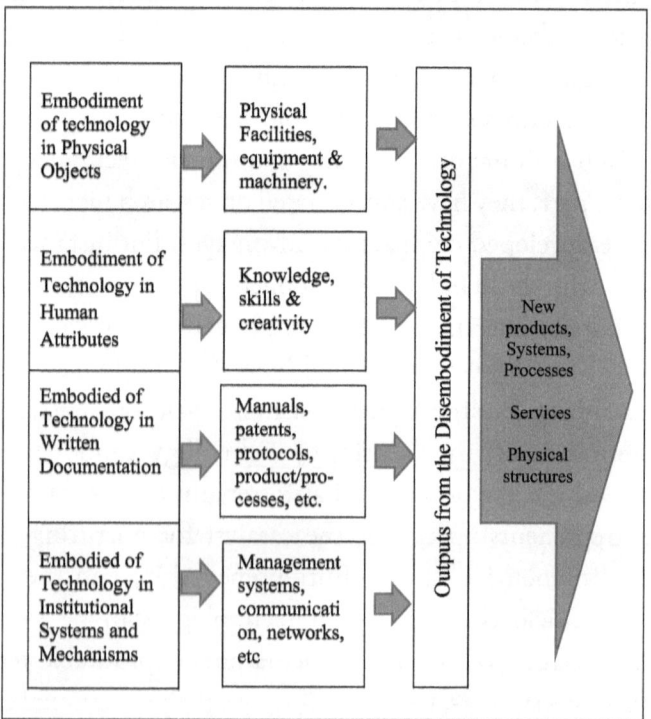

Figure 4.1: Embodiments Elements and Carriers of Technology

Technology Transfer Modalities

Before a specific method of technology transfer is selected, there should be a careful and detailed analysis of the technology, a strategy for future cooperation with the supplier of the technology, the technical capacity of the implementing agency as well as resources for further investment in similar technology or an upgrade of those that exist. Technology transfer modalities can be categorized as follows:

Foreign Direct Investment: FDI is considered one of the main channels by which technology is transferred, particularly at the state level. A foreign company generally invests in developing countries to create new markets, remove export barriers and get access to cheap labour. This type of transfer is done by multinational corporations that wholly own a subsidiary in a less developed country. The parent organization often provides capital, technology, management and marketing skills, while the foreign subsidiary provides some materials and labour. In this case, developing countries get all the benefits of technology transfer, particularly the development of their research environment as a vehicle towards the creation of new jobs. To attract foreign investors, the governments of developing countries generally make provisions for concessions in their policies. Without these concessions, large international corporations would not be motivated to provide long term investments in the country of interest.

Licensing Agreement: Licensing is an agreement under which the owner of a patent, trademark, manufacturers' rights or other intellectual property permits another company to use the technology developed by him/her in a certain area for a specific time. This enables the foreign firm to transfer technology to an affiliate or non-affiliate firm without incurring risk. When technology is transferred via licensing agreements between two affiliated parties, both direct foreign investment and licensing channels are employed.

There are two main types of licenses: (i) one which grants an exclusive right to use the technology; (ii) another with a non-exclusive right, which implies that the patent owner may transfer the right to use the technology to other companies in the same area. A licensing agreement could also include a sublicensing clause that permits the licensee to grant someone else the right to use the technology. As with any contractual arrangement, a licensing agreement should address all aspects of the relationship between the parties involved. This includes, for example, the currency by which royalty payments will be made and the means of determining the amounts of royalties. Most patent licensing agreements are concentrated in the automotive, chemicals, software, electrical and nonelectrical machinery, engineering, semiconductor, pharmaceutical and biotechnology sectors.

The advantage of buying a license/patent is that it can be obtained at a lower cost compared with other technology transfer methods. However, the purchase of a license requires sufficient knowledge, experience, relevant expertise and a manufacturing base for further in-house technology implementation.

In most cases, licensing products or brands is less expensive than establishing a wholly owned subsidiary in a foreign market. Licensing also provides the patent or copyright owner with the opportunity to expand into new markets with less investment. This is a means of displacing risks when the licensee covers local costs, and perhaps most importantly to capitalise on existing investments with little expenditure. Licensing also allows owners to capitalise on maturing brands in a home market and gain added revenue from other markets where the product or name brand may appear to be new or was not previously available.

Joint Venture: A joint venture is an agreement between two or more companies to execute a particular business venture. A joint venture arrangement implies that there are mutual assets, management, risks, profit sharing, co-production, services and marketing. In this channel of transfer, ownership of the venture in the less developed country is shared with a local partner. There are no fees charged for use of the technology. There are several benefits to be derived from a joint venture. Among the benefits are continued long-term cooperation between the parties, participants involved in the transfer are motivated to work on future ventures, and there is a lower cost of the venture being undertaken as a result of working together.

The disadvantages of a joint venture are often associated with the different visions and goals of both partners and their inability to be independent in management. As is often the case, companies are not always able to determine objectively the value of capital contributed by each party and therefore, subsequent profit distribution creates a problem. The foreign company often provides innovative technology and competent management, while the local company which is familiar with the market and related regulations provides the necessary investment and guidance.

Equipment Acquisition: Equipment acquisition is one of the most common methods of technology transfer. The main disadvantage of this method is that the company limits itself to mere technical knowledge incorporated into the equipment. The company does not receive any new management and production competence. There is no privilege in accessing any specific equipment since it is available for purchase by any other competitor.

Sub-contracting or Co-Production: In this case, the transferrer of the technology subcontracts (produces) part of the product components in the developing country, using inputs and technology supplied by the foreign company.

Turnkey Contract: This form of transfer allows the foreign company to take responsibility for establishing manufacturing plants in a developing country and sell as a package including all the know-how and skills necessary for plant operation by indigenous workers. The foreign firm retains no equity interest. In the case of a turnkey agreement, the general contractor is responsible for all the procedures related to technology transfer, including technology design, financing, equipment supply, construction and commissioning. The advantages of a turnkey agreement are that (i) the company concludes a contract only with one supplier who takes full responsibility for the project execution; (ii) except a force majeure, the project will have a fixed price; (iii) the supplier guarantees the performance and the efficiency of the technology.

A turnkey contract has distinct advantages, including (i) the company should know in advance all the features and output parameters that the technology should have after its launch; (ii) An in-depth knowledge of the corresponding field (in this case an independent expert organization could be employed to determine the technology's features and output characteristics) is required for a complex or large-scale technology; (iii) transfer price under a turnkey agreement is generally much higher than with any other method (the more risks the supplier takes, the higher the price); (iv) during the transfer implementation, a company doesn't have full control over the progress and quality of each stage of the transfer; (iv) contractor's financial problems may lead to the project suspension (it is difficult for a company to determine supplier's financial capacity and its ability to self-finance all stages of the transfer).

One of the ways to reduce the risks of a turnkey agreement is to involve the supplier in the capital of the new entity. This will motivate the supplier to ensure the quality of the new technology as well as to draw from the supplier's experience in further operational processes.

Foreign Company Acquisition: A company may acquire a foreign start-up company, which is pursuing the development of new technology. As a result, the company will not only get the technology, but also a team with the capability to develop it in the future. Moreover, the acquisition of a foreign firm automatically places the company on the international market. Among the main risks of buying an existing firm is the possibility that key employees may resign after the acquisition. Furthermore, the founders of the successful start-up would agree to sell it only for a price significantly higher than the market price. This increases the risk of profitability in the future.

Franchising: Franchising is an agreement where one company grants the right to use its trademark and business model to another company. The buyer of the franchise starts manufacturing and selling the goods according to the seller's specifications. Normally, the owner of a trademark also shares his/her experience in operating and managing the franchised product/technology. The main advantage of franchising is that the company gets an already-made brand. With the franchised product, the company acquires a proven business model and knowledge in management and marketing.

There are several disadvantages associated with franchising. Generally, the company is heavily dependent on the technology owner. In most cases, the company is required to purchase raw materials, equipment and other products from specific vendors. It is also required to follow the internal rules and procedures of the technology owner. The company cannot bring the product to other markets or sell the franchise. In addition, the decline of the franchise owner's reputation could have an impact on the company that has bought its franchise.

Buy-Back Contract: A buy-back contract is a form of agreement between developing countries and large foreign companies. Under this agreement, a foreign company supplies industrial equipment in exchange for profits derived from the sale of raw materials or goods produced on this equipment. This kind of technology transfer is often used in the construction of new plants in developing countries. In this case, the state becomes a shareholder in the created enterprise.

Through buy-back contracts, developing countries get an opportunity to acquire high-tech equipment from a developing country without directly investing in it. Moreover, the foreign company is responsible for the performance of supplied technologies. One of the potential disadvantages of a buy-back contract is that the foreign company is not motivated to start production at least costs which will certainly affect the quality of execution. Typically, under a buy-back agreement, the price for new technology is much higher than in the case of direct investments.

Original Equipment Manufacturer (OEM): OEM can be considered as a form of subcontracting, where a local firm starts manufacturing under the foreign company's specifications. Under this arrangement, a foreign company transfers a part of its technology and equipment to the local firm. It conducts training and management reorganization. The foreign company also sells goods produced through its channels and under its own trademark. OEM agreement enables local companies to absorb new technologies and reorganize their production. With new equipment and skills, these firms can produce new goods for the domestic market under their own brand. The main drawback of this agreement is that the local company is obligated to supply products at a fixed price to the foreign company, which is normally much lower than the market price.

Support Contract: In this agreement, the technology owner participates in the technology implementation by providing technical support at each stage of the transfer, technical support as well as personnel training. The involvement of technology developers in the transfer process ensures that there is closer cooperation between the parties which gives effect to the complete transfer of all knowledge and skills related to technology. In this way, the support contract may be a part of the licensing agreement to improve the transfer efficiency.

Non-Commercial Channel of Technology Transfer

The channels through which technology circulates are diverse and there is no best channel to be used for the successful transfer of technology because technology does not exist in a social vacuum, but is embodied in products, processes and people. The following are non-commercial channels to which technology can be transferred:

1. Exchange of information, international conferences, universities and research institutions, trade fairs, etc.
2. The use of foreign consultancies in technology-related activities.
3. Education and training.
4. Technical assistance: aid donors can provide experts at below-market rate to perform operational training or advisory roles incorporating technological training.
5. Management and service contracts.

Enabling Factors for Technology Transfer

The transfer of technical knowledge is "person embodied". This requires the mobility of people as well as personal contacts. The following factors are fundamentally critical to the successful transfer of technology:

(i) **Public Policies:** These policies can, to a large extent determine the success or failure of the transfer of technology process. Policies on matters pertaining to foreign exchange, taxes, lack incentives schemes, foreign investment legislation and indigenisation can influence the transfer mechanism in ways that make the transfer process extremely difficult or render the effort untenable.

(ii) **Education and Training:** Education and training is a prerequisite to the transfer of technical know-how. The education system should be so designed that it addresses the long-term needs of the country. Education is an important tool in building technical capacity particularly in science and technology and innovation at the national level so that the appropriate skillset necessary to support technology are identified for national development.

(iii) **Management Process:** The success in the application of technology is highly dependent on efficient technological management at various levels of national development. Successful management of technology depends to a large extent on an understanding of the technological requirements of a country or organization vis a vis manpower planning and the availability of the relevant tools for effective technological management. In many small developing states, there is a dearth of innovation-oriented management at various levels of the organization to effect proper planning, diagnose and solve problems of a technological nature. Effective management allows for the efficient use of limited resources and the management of various technologies.

(iv) **Resource Availability:** Resource endowment and availability often enhances the socio-economic prospects of a country and the lack thereof, inhibits a country's ability to diversify into other economic ventures which could invariably hinder growth and development. For the most part, economies that are primarily dependent on one resource run the risk of economic stagnation. Moreover, mismanagement and corruption of a country's resources could place that country at risk of experiencing negative growth. Countries such as Switzerland (which is land-locked), Japan and Taiwan with few natural resources, have made considerable economic strides and are now some of the world's leading industrial nations. The key to their success is not exploiting abundance resources, but creating it, not enjoying advantages, but coping with disadvantages. The long-term economic prosperity of a nation may not necessarily depend on the availability of resources but must recognise its limitations and devise innovative ways to make good of them.

(v) **Technological Constraints:** The needs of society as well as how technology can be used to address those needs should be at the forefront of the consciousness of decision-makers. This awareness should guide them in planning for the successful transfer of technology by providing the relevant support systems or services that are required. For instance, where the technology involves a high degree of risk to society, such as a nuclear plant, appropriate mechanisms should be developed to reduce such risks. Technology should be assessed based on its reliability, maintainability, efficiency, aesthetics and safety.

(vi) **Socio-Economic Development:** The impact of productivity on society should be an important consideration in the transfer of technology. Productivity can be generally affected by internal social and economic conditions in small developing countries. For example, an unskilled labour force and poor management skills could negatively impact the transfer process. On the other hand, successful implementation of technology can improve productivity and result in a higher standard of living.

(vii) **Cultural Value System:** Culture can be broadly defined as "a system of shared ideas; beliefs, knowledge, values, customs, traditions, symbols, behaviours and artefacts that belong to a certain society." There is no doubt that the transfer of technology has had a profound impact on societies at large. The greatest challenge facing society is the management of cultural change. The introduction of new technologies has engendered several changes that require adjustments to many facets of life including changes in education and training and the adoption of new livelihoods due to advanced technologies that are likely to affect the lives of the individuals and the general community. In every aspect of life, technology has radically transformed society in negative and positive ways. Information and telecommunication technology facilitated by the internet and the media have exposed various cultures to the world and have allowed people to learn the various cultural similarities so that they can be easily adapted by making the relevant adjustments. In doing so, physical barriers and borders are removed and people can interact across different cultural landscapes. Notwithstanding the enormous benefits of technological application, technology can create cultural imbalances. and social disruption. It is anticipated that artificial intelligence (AI) will revolutionize society in ways that were never envisaged.

Some technologies are complex and with high cost associated with them. In this regard, the ultimate success of the transfer process will to a large extent depend on the indigenous landscape and infrastructure, technical competence and institutional support of the recipient country.

Technology Appropriation

There is a view that technology can be transferred across frontiers in a whole scale manner without regard to the type and scale of operation, local conditions, cultural dynamics and other prevailing conditions that may impact the performance of the technology. The transfer of technology should be predicated on a set of techniques that make optimum use of available resources in a given environment. Technology should be most suitably adapted to the conditions of a defined situation. This is what is referred to as appropriate technology.

In transferring technology, consideration should be given to the human, environmental, and social conditions as well as the financial and material resource requirements that surround its application. "Appropriateness in technology can only be defined in relation to a specific setting or location. What is appropriate in one environment may not necessarily be suitable in another." Technology needs to be restructured to suit indigenous labour needs, climate, the socio-economic and cultural landscape. The design and scale of equipment and machinery may have to be restructured and retrofitted to local conditions and indigenous resources may have to be adapted to the imported know-how.

The management of technology should be wary of the medium to long-term effects of technology transfer on the wider community. Those who are disadvantaged by the adverse effects of a given technology may not be able to find immediate alternative employment. Therefore, they should be given prior consideration when determining the scale of the transfer of any new technology.

A small industry in a North African country was engaged in producing leather sandals for local use. As a precondition to receiving development assistance, it was suggested that by changing from the traditional leather material to plastic, a cheaper product could be made. Two plastic injection moulding machines were purchased at a cost of US $100,000. The highly automated plant operated successfully, but the net effect was that 5,000 leather shoemakers were put out of work and in turn, the incomes of the makers of leather, glue, thread, fabric linings, tacks, dressings, polishes, hand tools, wooden lasts and boxes, all of whose livelihood were connected to this industry were drastically reduced. In their place were just 40 injection moulding operatives. PVC plastics, spares, and technical expertise were needed from outside the country. The overall effect was a decline in domestic income, a continuing demand for foreign currency for the imports of materials and equipment and a declining standard of living in the area.

The Impact of Technology Transfer on Development

During the development process of a country, there would be a need to transfer technological know-how in the form of industrial plants and machinery, equipment and technical know-how to develop new products and services. Technology transfer can impact a recipient country negatively or positively. The positive impact can be experienced directly and indirectly at the micro and macro levels. Policies adopted by governments are critical in determining the extent to which technology transfer will be successful in achieving national goals.

The lack of technological capacity in several key sectors in many small developing states is in stark contrast to developed economies, which by far has outpaced the technological growth rate in developing societies. It is difficult for many developing countries to keep abreast of the technological advancements that are necessary to drive the economy and enable small states to compete. Technology transfer allows developing countries to increase their productivity through the use of sophisticated technology and innovation often associated with those used in developed countries to produce a range of goods and services and provide jobs that would not otherwise be provided. New technology which emanates from existing technology is easily adaptable and would often cause less social disruption. Appropriate technology is generally scaled to size in relation to several socio-cultural and economic realities. This allows enterprises to adjust their production systems to reflect the market demand.

The agriculture sector is an interesting case in point where the application of technology can have a significant multiplier effect. Agriculture, food and nutrition are so vital to small states that they represent one of the most valuable areas of potential application for new technology and innovation. Notwithstanding the land constraint that exists in some small states, every country has a responsibility to provide food for its citizens. Farmers will continue to apply nutrients to agricultural lands to ensure maximum crop yield. Hence, the application of pesticides, insecticides and fertilizer has become necessary. However, their persistence in the environment and their tendency to become concentrated in organisms high up in the food chain when injected has become a threat to many species of life including man.

The increasing world population growth has led to chronic food shortages and famine in some developing countries. In an age when there is heightened sensitivity to the destruction of marine life and its ecosystem, it is of paramount importance that various species of plant and marine life be preserved. Apart from the need for relevant laws, new technology in aquaculture for example, can help to preserve declining endangered species of fish, thereby preserving the fish stock as well as increasing the potential to earn substantial foreign exchange for countries where fishing is a key economic activity. The transfer of technology germane to the agriculture sector should be inevitably a viable option to address critical challenges facing the industry. For instance, biotechnology including genetic engineering has been used in developed countries to improve crop varieties as well as crop yield beyond current levels. It has also improved the nutritional values of crops, thereby reducing the growth cycle of some plants and permitting more harvests per year. New technology in agriculture can also increase plant density and strengthen their resistance to disease and climatic variations. Through technology transfer, hormones have been developed for crossbreeding and the conversion of animal feed to high animal protein, which enhances the quality of animal husbandry in developing states.

The intervention of new technology could auger well for many ailing traditional industries. Such transfer can significantly contribute to narrowing the gap between the rich and poor and increase efficiency, reduce waste and promote better utilisation of resources as a result of more knowledge and quicker industrial progress. There is a school of thought that, once the technology is foreign to a small state, it is perceived as being superior. Such thinking inhibits investment in indigenous R&D, which can result in repeated purchases of similar technology without proper assimilation or absorption.

Over the years, many small developing states have undergone comprehensive restructuring of their public and private sector enterprises, which has resulted in the redeployment of resources, both human and capital. In cases where such adjustments are designed to facilitate the adoption of new technologies, it will require a highly trained and skilled workforce. When a country experiences poor economic growth more often than not, it is the result of a drastic cut in funding for several key sectors such as education and health, which can impact the availability of training programmes, equipment and physical facilities. Consequently, these social measures can retard the development of a skilled workforce.

Unemployment is one of the most common socioeconomic problems in many small developing states today. Technology transfer is often regarded as a tool to bring economic progress and improved standard of living by addressing the unemployment problem. Paradoxically, sophisticated technology has its attendant challenges, particularly if it were not effectively managed.

The question of redundancy arises, particularly among the older workers who are more reluctant to be retrained for new opportunities in emerging sectors. This situation often results in a surplus of labour in some sectors and a shortage in others. When enterprises are unable to upgrade or modernise their operations to enable them to be more competitive, they have no other option than to go out of business.

Establishing the Framework for the Transfer of Technology

Successful transfer of technology should have an element of sustainability, both from the point of view of the environment and human resources. It is reasonable to argue that public policy should be instituted to guide the process of technology diffusion. Most importantly, critical issues such as employment legislation should be established to protect the interest of workers who are sometimes displaced as a result of redundancies or who in other ways find themselves at a disadvantage due to the transfer of new technologies. Retraining workers for new jobs may become the next best option. The costs associated with these exercises could be built into the transfer arrangements.

Public policy through incentives and disincentives should ensure that the importer and exporter of technology be held responsible for any damage to the environment. As a preventive measure, the new technology should have safeguards designed and control features, accident prevention capability and other contingency measures. Disposal of toxic waste and emission of toxic fumes, for instance, should be given priority attention. New technologies may require companies to develop health and safety schemes as well as insurance policies for workers. Existing policies that are inappropriate may need to be reviewed.

The policy framework for technology transfer should be based on an internal assessment of an organization with regard to what technology exists, what is required, and the most appropriate technology that is available on the market for any specific application. Some organizations practically try to keep abreast of the latest technology available. The introduction of new technology should be commensurate to the current level of operation and output, or it should be incrementally justifiable. In some cases, new technology is imported for the prestige associated with its novelty rather than the needs of an organization regarding its productive propensity. This approach should take priority when deciding on the implementation of such technology. It is worth mentioning that machines and equipment that are underutilized are tantamount to financial resources expended in an unprofitable venture.

The limited market in some small developing states coupled with fierce competition in the global market may not justify committing substantial scarce resources to extensive research and development programmes. It therefore seems feasible for governments in collaboration with R&D institutions to establish a priority of needs in relation to their geopolitical and socioeconomic realities. A balance should then be established between what technology should be imported and what should be developed locally. The decision to import technology should take into consideration such factors as cost differentiation, process modification and adaptability and technical expertise requirements.

It has been established earlier that for most developing states with a strong agricultural and agro-based infrastructural framework, appropriate technology research capability in agriculture is justifiable. However, it will be difficult to radically revolutionise the agriculture sector overnight until there is a social and political change that will facilitate the redistribution of income, review inequitable land ownership patterns, and reform credit systems that will provide support for small farmers and manufacturers.

Within the context of developing states, the transfer of technology should be carried out with various social and economic constraints in mind. Apart from the cost factor, which is a major determinant, the question of overproduction should also be given serious consideration. Hence, there should be checks and balances in the entire transfer mechanism. The production capacity must therefore be a function of the market absorbability to avoid a plethora of products in the production arena. If the need for new technology is not carefully assessed, new imported technology could be problematic. It is therefore important to note that output in excess of market demand will result in several functional equipment becoming idle and eventually made obsolete. In this regard, the cost attributed to the transfer of technology in absolute terms would by far outweigh the anticipated cost.

When technology is transferred across frontiers, it provides dual benefits to the importer as well as the exporter of such technology. It is for this reason that the importers of technology should take a holistic view of the social, cultural and economic impact of the technology. The importers of foreign technology should ensure that they can control the terms and conditions of such importation so that the prospects for technological development can be improved in the recipient country. National needs and regional priorities should be defined in terms of marketability and resource availability. The ability to bargain from a position of strength depends to a large extent on the information available on the technology. Hence, the technology should be properly researched, and all its facets thoroughly understood.

A small packaging company purchased some hardware and software from a firm in a developed country. Within two years the machinery began to malfunction resulting in the temporary closure of the plant. The management of the small company had to fly in expertise from the country where the technology was purchased to resolve the problem. This was a situation where there were poor technology transfer negotiations and prior contractual arrangements regarding training local staff for equipment maintenance, supply of spare parts and software update, among other things.

Quite often, agreements (between the importer and supplier of technology) that form the basis for the transfer make little provision for flexibility which can be disadvantageous to the importer. Sometimes technology supplied to small states fall short of the required standards. Against this background, there should be safeguards to ensure that technology provided to Third World countries meet the stipulated standards and can perform the task for which they were purchased. Third World buyers of technology are generally in a weak bargaining position because of their lack of expertise, lack of information or lack of alternative sources of supply, and therefore need professional advice in the purchase of new technology.

The technology transfer process should involve sequential phases for the implementation of the transfer to be successful. It would also require detailed planning from the identification of the technological need to the actual implementation and use. The following phases can be considered critical for the transfer process:

- Identify and evaluate the technological options including project feasibility.
- Determine the channels to be used in the transfer process.
- Adopt the given technology to meet the defined need.
- Integrate the technology into the company's overall operations.

Chapter 5

Commercial & Technological Innovation

There have been numerous definitions of innovation. Most of these definitions seem to suggest that the end product of an innovation process is a new tangible product of a technological nature. However, there is a contradiction to popular opinion that innovation does not necessarily depend on technology, but can occur entirely in the arts, education or in a social context. Innovation in the arts may involve pioneering new art forms, techniques and interpretations. There have been useful definitions that embrace the idea that innovation is more than a technological process. More importantly, that idea can be taken to the commercial level in recognition of a market need and developing a useful product, technique, or service to a point where it gains initial commercial acceptance.

It is necessary to define a few terms that would be used frequently throughout this discussion including organization, new product development and marketing and design. An organization can be regarded as an organised body of people or systems established to achieve a common set of prescribed goals or objectives. A product can be generally defined as a good, a service, or an idea. Goods are normally tangible items, something that can be touched and is physically divisible. On the other hand, a service is any act or performance through the application of human or mechanical effort that one party offers to another. An idea includes a concept, philosophy or notion. Both service and idea are intangible and are not physically quantifiable. In the context of this discussion, a product would be generally classified as tangible and intangible and further distinctions will be made whenever they become necessary.

In the Department of Trade and Industry 'White Paper' (1994) Phil Robinson craftily summed up the various definitions of innovation as the "successful exploitation of new ideas". Robinson's definition is not purely about products, it covers new ways of doing business with existing products or services and focuses on making money out of new ideas. This process includes marketing innovation (reinventing the product's image), innovation based on an unchanged product and changing the corporate strategy, particularly the marketing strategy. Brian Twist (1986) noted that "Innovation cannot be divorced from creativity. Successful innovation is offering the market something new for which the customer is prepared to pay. This arises from new technology or a new application to existing technology."

The process of innovation can begin serendipitously and ends up in commercial and profitable products. Innovation is an important ingredient in the creation of value for an organization. It is considered a key driver of growth (through the introduction of new products and services) and efficiency (through process improvements) in organizations. Innovation can bolster new product development by identifying areas for improvement in existing products in targeting new market opportunities.

The Innovation Process

Table 5.1: Typical Stages of the Innovation Process

Idea Generation (Preconception Period)	Innovation Stage	Post Innovation Period
• Screening • Discovery	• Experimentation • Invention • Application	• Diffusion • New product development • Product improvement • Commercialisation

The complexity of an innovation process explains the long-term gestation period for product development. It is estimated that a major innovation involving basic research requires more than 20 years. A study was conducted on ten major innovations including a heart pacemaker, hybrid grains, electrophotography, oral contraception, videotape recorder, etc. It was found that the timeframe between conceptualization and product development varies from 6-32 years. Small companies that are willing to innovate stand a good chance of navigating their way through the vagaries of a dynamic competitive market and becoming competitive and successful.

Technological innovation can be categorized as (i) Product Innovation and (ii) Process Innovation. Product innovation involves the development of new products or the improvement of existing ones. Similarly, process innovation relates to new or improved production processes. These classifications do not suggest any clear distinction between the groups. The deciding factor lies in the degree of novelty and change involved. It must be noted that the innovation process begins with the creation of ideas. Ideas can originate from several sources including various departments such as production, marketing and management. Sources of ideas outside the company could include suppliers, customers and patents from suppliers. Ideas should always be encouraged, and there should be a process of screening ideas to separate those that have value from those that do not offer the potential for creativity. Idea generation is a result of creativity and there can be no innovation without creativity.

Success in product development begins with creativity. It refers to that thinking which results in the generation of ideas that are both novel and worthwhile, or ideas that can further add value to what exists. Imaginative ideas are insufficient because industry primarily needs practical ideas. Ideas are generally generated from a specific market need or from the R&D department. Hence, creative thinking should be converted into innovative actions. Creativity and good ideas are all encompassing. It encompasses research activities, the development of new products, the creation of consumer markets, the delivery of services and a myriad of interrelated phenomena that operate in an integrated manner.

The process of innovation involves grappling with unknowns or uncertainties that may be technological, economic or merely a manifestation of a creative process or a product not hitherto known. The process is an informational one, and an inordinate amount of time must be spent obtaining and using information about every conceivable issue to be addressed. The process of innovation can only be regarded as successful if the end result was offering the market a product for which it is prepared to pay. Sometimes the innovation can be a new technology or an application to an existing technology.

Creativity is the ability to generate novel and useful ideas about products, services and practices. Innovation, on the other hand, is the successful implementation of creative ideas. New ideas are often filtered, developed and commercialised into useful products. Creativity drives innovation and innovation is the key to sustained competitive advantage for an organization in times of intense competition. The success of an organization depends on its ability to continuously innovate in a manner that a firm can churn out new or redesigned products to the market at a faster rate than its competitors. The innovation process can be initiated through several techniques that can stimulate creativity (Fig. 5.1).

Brainstorming is one of the techniques that will enable an organization to innovate and grow through the creation of new ideas. At the employee level, creativity results from a combination of expertise, motivation, and thinking skills. At the team level, it results from synergies among team members greater than the sum of its individual parts. Through brainstorming, new ideas are shared that give rise to a chain of new thoughts.

During a brainstorming session, all ideas emanating from within the group should be welcomed so that participants are encouraged to keep generating new ideas. Moreover, the group should be open to new ideas since several possibilities may emerge from these ideas. As ideas are filtered out, the emphasis should be on quality and not necessarily on the number of ideas being generated. Creativity is a process that allows one to draw from one's imaginative ideas.

The 'Suggestion System' is an age-old method used to generate ideas. This approach requires greater involvement of employees from all levels of the organization. A suggestion system should be integrated into the organization's culture of employee participation. The organization should provide an environment conducive to free submission of ideas. Information leading to technological innovations can also be generated from the following sources:

- Feedback from dealers, installers and customers
- Reverse engineering
- Patents
- Exchange of information between friends

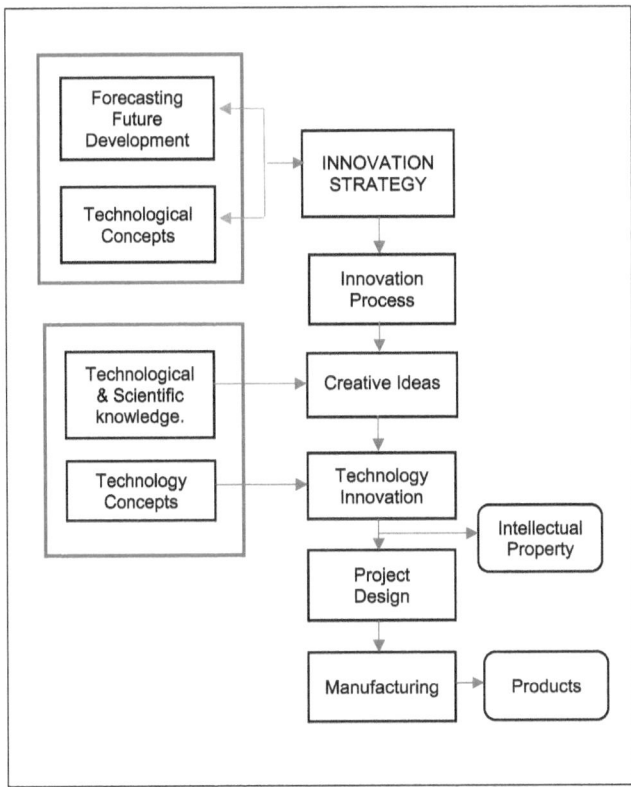

Figure 5.1; Depiction of the Innovation Process

Before the process of innovation can produce tangible results, several barriers must be overcome. Among them are:

- Poor management;
- Marketing impediments;
- Lack of capital, particularly venture capital for high technology businesses;
- Lack of government policies relating to patents, anti-trust and other regulations;
- Misdirected education and training;
- Weak technical infrastructure in communication, electricity, science and technology information bureaucracies.

The story of SM Jaleel is not only inspiring but demonstrates how an enterprise in a small developing state can become a world class entity through strategic thinking driven by technological innovation which engendered new and improved product development. SM Jaleel is a soft drink manufacturing corporation in Trinidad and Tobago. It has gained a significant share of the CARICOM market through its four brands. The company has been in operation for more than 80 years but has only risen to prominence over the last 25 years. Its success began with the change of management which transformed the enterprise into a world class company through strategic thinking driven by technological innovation.

The company invested heavily in state of the art plant and machinery and combined a modern integrated production process with a strict quality control system. Moreover, it has also made a significant investment in its human resource through regular training programmes for its employees. It also combined classroom training sessions with on-the-job training. This investment has resulted in increased labour productivity and staff retention rates. The products are sold in many countries around the world.

This case typifies many of the issues that were discussed earlier and others that will be discussed in subsequent chapters. The following 5 pillars can be considered key inputs to the success of any company:

(i) Technological innovation;

(ii)   Strategic marketing;
(iii) Human resource development;
(iv)   Quality driven;
(v)   A strategic vision to become world-class.

SM Jaleel has methodologically implemented many of the approaches and techniques that have been discussed earlier. The principle exemplified was to find a need and fill it. They studied the demands, challenges and opportunities of the global market and what was necessary to make them a world-class company. After two years of extensive market research, they identified an untapped market segment in 4-10-year-old kids who will be fascinated with soft drinks in a Chubby bright coloured 250ml bottle.

Strategic Approach to Innovation

Conceptually, a strategy is a plan of action designed to gain a competitive advantage over other competitors within a given market. "A corporate strategy is a framework that gives an organization an overall direction and impels it to take action." Strategies should be aligned to the organization's objectives and priorities upon which resources are expended. Organizations design their strategies among diverse groups within the organization and outline how various functions including marketing, finance and R&D will support those strategies. Planning and control of all activities of an organization is one of the functions of professional management. This function is critical because all activities can be coordinated in a manner that allows the organization to work towards a common set of stated objectives against which it can measure performance. This approach does not only apply to major organizations but also to SMEs. Similarly, an organization should develop its innovation policy objectives and then determine what strategies it hopes to pursue. An innovation strategy should therefore consider the following:

(i)   What are the key objectives of the innovation process that will form the central theme of the strategy? E.g., the introduction of new products, a new method of production or the opening of new markets.

(ii) Determine the innovation strategy to be adopted? (Is it "technology push", or "market pull," The term 'Market Pull' refers to an expressed need for a new product or a solution to an existing problem regarding a product, which comes from the marketplace. The need is often identified by potential customers or market research. A product or a range of products is developed to address the original need. "Market pull" sometimes starts with potential customers asking for improvements to existing products. Focus groups are often central to this approach when testing a concept design or an existing product. "Technology Push" on the other hand, occurs when R&D or a technology breakthrough drives the development of a new product. In the "technology push" approach, the focus is on technical issues and problem-solving. This triggers a search for scientific knowledge both within the firm and from external sources. The firm then develops an innovative technical solution and offers it to the market.

(iii) In what form should the innovation be initiated? (incremental innovation, radical innovation, or disruptive innovation?)

The following factors should therefore inform an innovation strategy:

1. The company's capacity to innovate;
2. The cooperation between departments within the organization;
3. Strategic intelligence;
4. Market dynamics
5. Competitors;
6. Legislation to facilitate innovation.

Before a company can develop a strategy, it should first develop a vision about where the company sees itself as a competitor in the market. It should then delineate the objectives that should be achieved to position itself in the market as a major player. Similarly, the enterprise should develop an innovation strategy, which is a component of the broad corporate strategy. The innovation strategy should provide a sense of direction regarding policies, actions to be taken, and how resources are matched to achieve organizational goals.

Decisions made during the planning stage should take cognisance of the future of the business environment in which the company will be operating. A contingency plan should also be drawn up to reflect any changes in circumstances from the time when the plan was drawn up to the current situation. It would be useful to examine forecasting techniques to determine trends in the market and the business environment, and to identify those factors that may present threats to existing businesses or those that will offer opportunities for new business ventures. This information may consequently help to shape strategies.

In the long-term interest of an organization, one of the goals should be to create an enabling environment to foster and promote innovation. During the planning process, the organization should clearly define what the innovation targets are and how they can be sustained by resources, processes, technologies and behaviour within the firm. A firm oriented towards innovation should include these targets in its strategic planning as they would be reflected by its culture. An innovative company generally develops the infrastructural framework and processes from which it sources solutions to problems in product concepts and designs. Strategies should be promulgated at all levels throughout the organization through policy statements and action plans.

For commercial innovation to be successful, the emphasis should therefore be on product development and marketing innovation including improving the product's image, repositioning the product in the marketplace, as well as differentiating it from its competitors to gain market share and competitive advantage.

Managing the Innovation Process

The outcome of the innovation process cannot be guaranteed. However, the success of innovation relies heavily on the continuous interaction between marketing and product design. Innovation may be derived from a new or old technique, or it may emerge from the result of research or the solution to a technological problem. At some point, there should be an interaction with the end users regarding the outcome of the innovation process. R&D, which is somewhat a complex process, relies on frequent interactions between research, technology and markets.

On the one hand, innovation represents a way of developing the business and establishing the environment to survive in the face of fierce competition. On the other hand, innovation is a risky process, but it is also a necessary risk, as Brian C. Twiss wrote, "Successful businesses of the future will be those that learn to live with uncertainty and incorporate it into the decision-making processes." Many studies carried out referring to innovation have demonstrated that by applying management methods and techniques to the innovation process the risk of failure would be considerably reduced.

In the realm of manufacturing, marketing information should transcend those boundaries. Although marketing is charged with generating new product ideas, some companies use product development teams to evaluate competing new product development and major modifications of current products as well as to develop detailed designs and specifications.

New product development describes the process that transforms technical ideas or market needs and opportunities into new products. It differs from technological innovation in that, a new product may involve changes in form, components, materials or even packaging. Modern marketing makes it possible to commission a product positioning study, which may reveal the repositioning of the existing products in the market. By tinkering with the elements of the marketing mix, new and improved products can reach the supermarket shelf promptly. Products that are not necessarily new or improved can be repositioned in the consumer's mind as 'new' with astute use of the media and creative packaging.

Firms can also opt to expend resources on product modifications, which are less costly and risky in the short term. Although considerable profit can be made, in a crowded market, this situation would not prevail for any long period. The question of why a consumer should buy a product in preference to another comes back to innovation and product design for which the customer is willing to pay.

Product innovation plays a key role in the growth process of a firm. Empirical studies have shown a nexus between new product development and business competitiveness. Innovation manifests itself in new products that represent its pervasiveness and a driving force in high-growth firms compared with low-growth firms. Growth is interwoven in a company's philosophy and its drive to be world-class.

It is integrally bound up with a firm's culture and ability to innovate. The sustainability of growth potential is not only related to a firm's desire to forge ahead of its competitors but also its ability to bring products to the market quickly because of its superior innovative capacity. Continuous innovation gives rise to inventiveness and provides leadership in the market.

In an era when the focus is heavily on developing a knowledge-based economy, open network-based modalities can allow SMEs to access the source of innovation thereby enabling them to participate and contribute to the innovation process. Such opportunity can be influenced by the spillover of knowledge into new areas of interest that allow SMEs to access networks in a manner that will increase their opportunities to partner with larger firms. This can be realised through efforts of globalization which have facilitated cross-border collaboration in innovation through which inputs in the form of ideas, finance and technical skills can be obtained from foreign countries. On the other hand, outputs such as products and services, patents and licenses can be exploited from foreign markets.

Innovation and Marketing in Product Development

New product development is often considered the lifeblood and hope for the future of any organization. The diversity of complex products, customer idiosyncrasies, high volume of entrepreneurial risk and various interacting and influential marketing factors characterize an unprecedented era of intense competition in the market. These conditions have undoubtedly fostered an innovative marketing approach, which should prioritize the organization's concerns. The marketing approach and the factors that characterize this approach will also be briefly examined to give credence to the concept and reality of commercial innovation.

Experts in the field of marketing have defined the term "Marketing" in various ways. However, an analysis of many of these definitions reveals a commonality of key concepts such as exchange, needs, values and satisfaction. The concept of marketing is a function consisting of individual and organizational activities that facilitate the exchange of relationships. These relationships are geared towards the satisfaction of needs in an expeditious way and in a dynamic environment. Such needs could be satisfied through the creation, distribution, promotion, and pricing of goods, services and ideas.

Every individual has a set of needs or requirements that are articulated in statements or phrases which are biologically and physiologically inherent. It is incumbent on an organization to design a product to meet those needs. This must be done through the product, which is being offered to the customers, bearing in mind that customers would only buy products for the value they provide to them. For example, if a product is perceived to enhance status, self-image, or sex appeal, it will appeal to customers whose perceived values are compatible with the inherent values of the product that is intended to provide the desired satisfaction.

The value of a product should be communicated through its physical features, including its design, uniqueness, ease of maneuver, etc. Customers' concern for value generally reflect the trade-off between the benefits of a product and the price paid for it. Moreover, it must be stressed that one of the principles of effective marketing is to get the right product to the right customer at the right time, at the right place, in the right quantity as well as at the right price.

The addition of value to a product will increase either the product's price or value. For example, offering one year of free support on a new computer would be a value-added feature. Individuals can also add value to services they perform, such as bringing advanced skills into the workforce. In today's marketplace, consumers now have access to a whole range of products and services when and where they want them and at the right price. For this reason, companies constantly struggle to find competitive advantages over one another. Companies should understand what value customers place on a product. The process in which a product is produced, packaged, marketed and how it is delivered to the final customer is a manifestation of the understanding (from the company's point of view) of the value customers place on that product.

When a BMW rolls off the assembly line, it is sold for a much higher price than the cost of production because of its reputation for stellar performance and sturdy mechanics. Additional advantages have been created through the brand and many years of refinement. Similarly, Amazon has been a force in the e-retail sector with its automatic refunds for poor service, free shipping, and price guarantees on pre-ordered items. Consumers have become so accustomed to its service that they are willing to pay for Amazon Prime memberships because they value the services that are associated with the purchase of its products.

The world of business has progressed in a very dynamic manner. Customers today are demanding products that are imbued with various levels of sophistication and innovation, yet with some simplicity of operation. In the corporate world, a progressive approach to product development is that the organization should develop a venture team. This team should comprise of representatives from various functional areas within the organization who are mandated to contribute to the development of designated projects The team should not be constrained by resources and regulations that will inhibit its decision making but must be given the free reign to make decisions including taking calculated risks.

It must be recognised that marketing has an important, albeit very specialized and critical role to play in the development of a product until it gets to the final customer. Ingenious design by itself is not enough to bring about commercial success for any product. Effective marketing is just as important in achieving success, irrespective of the quality of the design.

Marketing; A Driver of Product Innovation

Following the early years of the Industrial Revolution, many of the world's great products that entered the market originated from scientific or technological breakthroughs through R & D. Hence, there was a technological "push" rather than a market "pull" orientation to new product development. The subsequent failure of some products in the marketplace decisively forced technologists to redress this thinking. During the mid-1950s the idea of the marketing concept emerged to prominence amongst organizations that redefined their role from just making a product to one of determining the needs of the customers and then developing the product to satisfy those needs. The tenet of the "Marketing Concept" is to satisfy the needs of the customer, through a set of coordinated activities in concert with the organization's goals.

After many years, there is new thinking regarding the marketing concept philosophy. It is being argued that if, as suggested by the marketing concept, the firm should be consumer-oriented and that products should be conceived and introduced to meet the needs of buyers, then buyers' needs and wants must be identified, qualified and quantified at the conception of the product idea, prior to product development. The viewpoint is being advanced that a 'market pull' approach to product innovation is the logical outcome of implementing the marketing concept to product development. The argument is further refuted on the grounds that apart from some mundane products such as potato chips, feminine hygiene and deodorant; breakthrough innovations throughout history have been the result of technological breakthroughs, laboratory discoveries or an invention. In several cases, those persons at the forefront of these inventions are far removed from the customers or have a vague notion of the needs of the market. The recipe for success is "Find a need, then fill it." In effect, the market concept is encapsulated in those six words.

The emphasis should always be on the customer as the most important element around which all activities would revolve from beginning to end. The business should endeavour to keep its customers happy, by modifying and redesigning the product to meet their changing desires and preferences. Customer satisfaction could also be attributed to successful marketing (beginning with the design of the product) which involves listening to consumers' feedback through focal groups, one-to-one qualitative interviews and affinity diagrams.

Many players external to a company are part of the innovation process through strategic partnerships. Most firms maintain a close relationship with their clients and use innovation as a tool to respond to their specific needs. Clients take an active part in the innovation process by providing information and feedback that could lead to new products or enhancement of those that exist. This can be done through incremental innovative improvements on the product and in the production methods. Stakeholders including distributors, suppliers and the company's parent groups can also be actively involved in the process.

How Product Design Reflects Consumer Needs

It was earlier established that the customer should occupy the focal point of the marketing concept. Some products fail to attract the attention of customers because the firm may have ignored this most vital and decisive philosophy. On the contrary, customers' needs were often based on assumptions. According to the traditional approach, the first step in successfully implementing a customer-oriented approach to marketing is to conduct extensive market research. This means that the data collected should be carefully analyzed so that decisions can be based on accurate and reliable information.

Once customers' needs are clearly defined and understood, the firm should then proceed to develop the product that will meet those needs. This process does not only involve manufacturing the product, but the marketing department must also have an understanding of the intricacies of the competitive market. Competition often drives innovation, which ultimately influences change. In the entire process, marketing should be the catalyst for innovative change which is based on the needs of the customers.

The traditional approach to product development advocates

that the firm should make what customers want to buy rather than persuading them to buy what it makes. Customer satisfaction occurs when perceived performance matches and exceeds expectations. Hence, success in marketing lies in an organization's ability to exceed the value offered by its competitors. It is paramount to note that the trade-off for meeting customers' needs is the retention of current customers and the attraction of new ones. The retention of customers is vital for the future of the business. Fig.5.2 depicts three major functions (marketing research, product development and manufacturing) that are involved in the manufacture of a new product, and how they are interrelated.

For many companies, the adoption of the marketing approach has resulted in a preoccupation with minor changes to existing products, restyling exercises and trivial innovation. Breakthrough innovations like the telephone and electric light bulb are the result of 'technology push'. New product ideas and consequent modifications to those ideas do not remain with the design team. At the initial stage, the focus should also be on design for the customer and design should be informed by market intelligence. There should also be input from several departments that should take responsibility for the final product. As a result, a direct link should be created between design and manufacture. Several types of design considerations should be considered. For instance, the firm would be concerned about design factors such as manufacturability, frequency of design changes and opportunity for product design changes and the application of concurrent engineering practices.

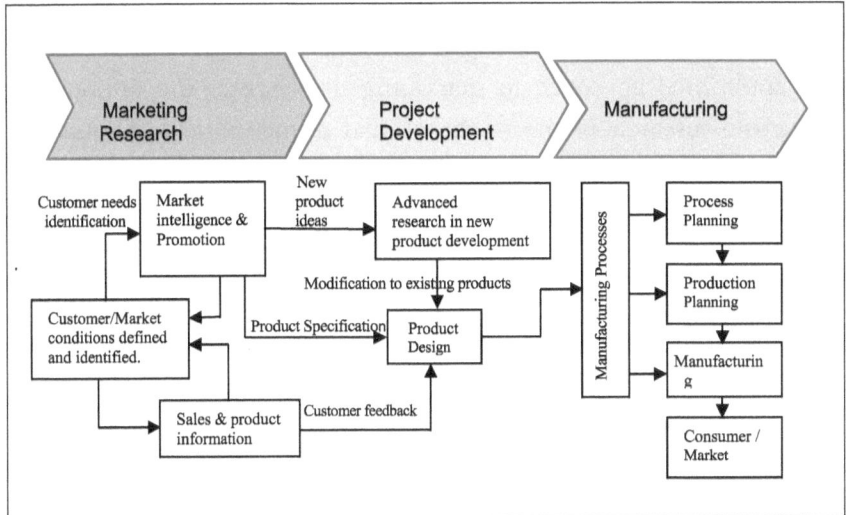

Figure 5.2: Flow of Activities in Product Design and Process Selection

The design also needs to match customers' requirements in product performance e.g., fitness for purpose, quality, durability, appearance, and price. Better design can improve operational efficiencies in manufacturing, distribution and services. Furthermore, product development is a cross-functional process geared towards the improvement of the product and the reduction of functional barriers. It is important to harness other skills to achieve successful innovation and to lessen the reliance on R&D. Among the skills necessary are marketing and design, which should be treated as a priority. The marriage of marketing and design, rational analysis and practical imagination often leads to dynamic product development. In Britain, the Department of Trade and Industry in its 1995 White Paper, noted that "The effective use of design is fundamental to the creation of innovative products, processes and services. Good design can significantly add value to products, lead to growth in sales and enables both the exploitation of new markets and the consolidation of existing ones."

How Marketing Information Enhances Innovative Design

The traditional approach to marketing underscores the notion that marketing research is one of the critical elements in developing an effective marketing database. Marketing research can be described as the objective identification, collection, analysis and dissemination of information for use in decision making. A well developed and effective information system will facilitate proper marketing analysis, planning, implementation and control of the entire marketing process. Information can originate both internally and externally and should be relevant to price, advertising, sales, competition, distribution and expenditure. Information should also be regularly processed so that it becomes part of the marketing data bank. Communication is undoubtedly one of the most vital links in every facet of life, and its role in marketing is no exception.

The importance of market research in product development has always been a contentious issue. This has led some prominent innovators to disclaim its importance. Christopher Lorenz (1993) in his article "Gut Feel is Market Research Too," has identified two approaches in assessing the potential market for innovative products.

His first approach is 'gut feel' which is moving beyond conventional trends and popular opinion and looking into the future for lifestyle changes. For instance, Sir Clive Sinclair sensed the potential demand for a low-cost home computer. Sony Chairman Akio defied the market research evidence, with the belief that the next generation of young people in search of entertainment will need a Walkman personal stereo. These highly successful products were a result of gut feelings and no market research, which undoubtedly has a role to play in the development of new products.

The second approach is qualitative market research or creative thinking, which Lorenz (1993) describes as an amalgam of fact-finding and social forecasting. This is in contrast to the conventional quantitative survey methods. Creative marketing is based partly on attempts to anticipate long term trends in lifestyle fashion and technology. A company needs to engage its clients, designers, and the marketing staff in meaningful discussions aimed at identifying problems with existing products and incorporating features that are compatible with emerging trends and what customers need. Radical innovations remain a risky business, but people who are risk averse easily gravitate to the safer route of "market pull'. A larger number of businesses are likely to survive if there is an understanding of the market, the customers, the business environment and the requisite linkages that need to be created among them.

The Role of Strategic Planning in Marketing Innovation

No business can survive into the future on short-term plans. Management should develop long term strategic plans to cope with unanticipated occurrences in the market. Strategic marketing analysis cannot be carried out in isolation from the wider process of strategic analysis. Notwithstanding its interdependence, strategic marketing analysis forms the most essential input in strategic analysis. Strategic planning encourages an investment portfolio approach to management. In this approach, the investment climate is constantly being monitored to decide on new investment thrust or reduction of capital projects.

An organization may also develop reliable mechanisms for potential investment that will enable the business to decide when to enter a market that is profitable or to exit an unprofitable one. Having identified its competitive advantage, it can then formulate goals and plan programmes and budgets to capitalise on its competitive advantage.

A new product strategy is part of the overall corporate strategy of a business entity. The basic strategic decision an organization can choose is whether to be reactive or proactive. A reactive product strategy deals with market challenges and pressures whenever they occur. On the other hand, a proactive strategic approach will require a firm to plan ahead and allocate resources for future events. A firm that adopts a reactive approach copies successful projects from the market; while in the case of a proactive strategy, it is the first to launch a product on the market. Proactive firms take several initiatives that will allow them to remain competitive and market leaders.

It is worth mentioning that competition based on cost could be a dangerous strategy in business since those firms with the largest budget are likely to absorb a loss and forge ahead. The story of Walkerswood, a Jamaican food company is very instructive in highlighting critical areas of strategic planning and marketing. Although many Jamaicans enjoyed jerk pork, they often found the spice too strong, hence, bottled seasoning would allow them to jerk as they choose. Early in the 1980s, Walkerswood a Jamaican cooperative became the first business enterprise in Jamaica to bottle the jerk mixture that was ready-made to season the pork they had been selling.

Walkerswood's recipe for success came from its effort to differentiate its product through the use of local materials and brand development. The company capitalized on the demand for authentic ethnic foods in overseas markets by providing value-added food products of higher quality than their competitors. Many of their competitors sourced cheaper ingredients.

Given its decision to compete on quality, which the company defines as authentic Jamaican taste. It was therefore critical that it relies on a consistent supply of local raw materials. Walkerswood was almost forced out of business when successive droughts in 1996 and 1997 caused the price of locally grown escallion, a key ingredient in its products to jump from single digits to up to J$100 a pound. The company decided to import cheaper scallion as a result. The quality of the imported scallion was not on par with that grown locally, hence, the quality of the final product was compromised to the point where sales fell.

The company then reverted to the purchase of locally grown scallion (throughout the drought) which guaranteed a higher quality but at a higher market price. This decision realised an increase in profit and ensured that Walkerswood recaptured market share and provided the customer with the level of satisfaction that they have grown accustomed to.

To ensure consistency in the supply of raw materials the company began a process of backward integration, which included buying a farm in that part of Jamaica that was least prone to drought. It also negotiated the rehabilitation and transfer of mined-out bauxite lands to its community of farmers. To further increase yield and improve the quality of its ingredients, Walkerswood provided irrigation and other technological applications to local suppliers.

Trading under the name of Walkerswood Caribbean Foods Limited the company has become a successful export-oriented food company which has captured market shares in the ethnic foods market and is gradually expanding into the broader crossover markets. They have also employed many Jamaicans in various areas of their business.

Walkerswood found out that compromising quality for cost can have negative consequences as it experienced the loss of customers and by extension its profit margin. It must be reiterated that quality should never be compromised for other factors of less importance. The marketing of the jerk spices targeted the Jamaican diaspora in the USA, where Walkerswood had an advantage. Importantly, the company learned several vital lessons.

One of the lessons learnt is that if a company wants to influence the quality of raw material that goes into the production of the product, it can do so in two ways (i) It can employ other small companies to undertake the production of the primary raw materials with an agreement to provide the necessary technical support to produce those raw materials. (ii) It can establish its own production facility (in this case a farm) where the primary raw material is grown using the requisite technology that will result in optimum yield and the desired quality.

Science & Technology and the Evolution of Product Development

An appreciation for scientific and technological issues is largely dependent on the structure and content of the education system in small developing island states and the popularization of these disciplines. The education system should be structured in a manner that it can respond to the demands and realities of industrial development in the twenty-first century and beyond. Every student should have an opportunity to study science and technology-related subjects.

Students should be encouraged to experiment, discover and create rather than to assimilate and recall facts and information. It is important to recognize that science and technology is a cross-cutting tool that can be used to improve efficiency, quality and productivity. Science and technology should therefore be seen as a tool that seeks to address critical development issues of small states.

Creativity is a function of the thought process which is borne out of a system that encourages people to think critically. Innovation, which is an embodiment of creative ideas, can be a major driver of the economy, particularly when it leads to new product development which has the propensity to yield high levels of profitability or to increase productivity. Although innovation may emerge because of less formal on-the-job modifications, an environment conducive to R&D can give rise to breakthrough innovation. Innovation can enable firms to use technology to develop cutting edge processes and enhance product development thereby maintaining a competitive advantage.

Science and Technology should be used discretely to address such issues as product development, biotechnology initiatives, and the services industry using computer and digital technology transfer for various applications. Appropriate technology can be aptly used for rural development, alternative energy, new agricultural technology aimed at increasing crop yield and technology assessment for economic planning. Technological awareness and the ability to understand technological possibilities and limitations are important considerations for decision-makers.

Many small enterprises are devoid of working capital and do not have access to credit, which will allow them to acquire the financial resources to invest in new processes, improve the quality of human resources and pursue technological innovation. Without the latter, there could hardly be any long-term strategic vision that speaks to innovativeness and creativity. Small states need to find the resources to invest in basic R&D and establish strong linkages with academia that will enable the result of research to be translated into marketable products. Several policy initiatives are critical in driving the process of innovation, including incentives to encourage creativity and ingenuity beginning in the school system from the outputs of science fairs, technology shows and the like.

Appropriate science and technology policy should be developed to provide a framework to guide the choice and application of science and technology in schools for national development in several critical sectors. Moreover, there should be strategic programmes to popularize science and technology to demystify the cultural phobia surrounding the subject. It should therefore be integrated into the cultural heritage of the country as a tool that will address problems in engineering and manufacturing. There is a need to establish National Innovation Systems to support the transition to an innovation-driven economy, through which public and private sector firms, universities and government agencies can interact to initiate the development and diffusion of technology. This approach will indeed form the basis for the innovation process that can lead to the development of products with commercial potential. It is within this context that the ideal opportunity will be provided for small developing states to establish the environment and framework for problem-solving and encourage lateral thinking to flourish, thereby facilitating the development of creative minds that will engender technological development.

Chapter 6

Traditional Development Framework for Enterprises

In most small developing economies, the private sector accounts directly for about 85 to 90 percent of GDP, with the remaining contribution from the public sector, which in turn is financed by taxes that are derived from private sector growth. Rising standards of living essentially depend on how well the private sector in most developing countries performs. Private sector performance also depends on how well a country promotes the development of its enterprises and the rate at which these enterprises can grow and become more efficient, and how well they can compete in international markets.

The development of a strong private sector is a complex process since there are many extraneous factors impacting the success or failure of the sector. Efficient and competitive private sector firms do not always develop solely because of their internal capabilities. Innovative entrepreneurs, skilled managers, and a dedicated and well-trained workforce are critical success factors of a firm and the extent to which it can contribute to the growth of the economy. Public sector policies and variables in external markets can significantly impact how well firms perform. Generally, there needs to be an overall enabling environment to promote private sector growth. The literature on private sector development in most small developing states suggests that the sector is in a state of flux and requires an injection of capital as well as an infusion of approaches and tools to enhance its competitiveness.

Enterprises in developing economies are generally categorized as micro, small and medium-sized in several areas such as agro-processing, food processing as well as some relatively small traditional activities including fisheries, arts and craft and construction. Over many years, several attempts have been made to group enterprises based on the number of employees, total net assets, sales and investment. The definition of SMEs has been the subject of considerable debate.

It is important to state that there is no standard international definition for small and medium-sized enterprises (SMEs). SMEs are defined differently in the legislation across countries, particularly so, because the dimension "small" and "medium" of a firm are relative to the size of the domestic economy. The criteria for an SME vary from country to country as well as the source for reporting SME statistics. There is no uniformity across countries on any given criterion to define an SME. For example, a large enterprise in Germany may be considered small in most of the small states in the Caribbean Asia and Africa. Many countries use employment, sales, or capital investment as the main criteria to define an SME. Against this background, there is no common definition of an SME. The following broad categories of SMEs seem to apply in most developing economies:

Micro enterprises: employment level below 10
Small enterprises: employment level from 10 to 49
Medium enterprises: employment level from 50 to 249

The SME department of the World Bank which works with many developing states uses the following definitions:

A micro enterprise up to 10 employees,
Total assets of up to US $10,000
Total annual sales of up to US $100,000

Small enterprises with up to 50 employees,
Total assets and total sales of up to US $3 million

Medium-sized enterprise with up to 300 employees,
Total assets and total sales of up to US $15 million.

It is inconceivable to think of a medium-sized enterprise in a small island state of about 350 square kilometers and a population of 120,000 people to employ 300 persons with assets and sales of US $15 million. This is contrary to the World Bank threshold of a medium-sized enterprise. The industrial group classification should vary from one country to another taking into consideration the country's size, population, resources and other development criteria. Failing to do so, one would run the risk of applying development indices in the wrong context and using comparisons disproportionately. When development agencies provide loans or grant support to enterprises in a particular geographic region, for example, in Latin America and the Caribbean, the disparity will then become obvious, because although Latin American countries are relatively poor, they are generally larger in size and their development indices are different, and they are at varying stage of economic development to small Caribbean states. Hence, small states of the Caribbean when lumped together in a regional context will be at a serious disadvantage.

It is becoming extremely difficult for small business entities to survive in the face of extreme competition from larger corporate bodies that have more resources. Small enterprises continue to encounter numerous drawbacks in their routine functions. Among them are low technology, lack of adherence to standards and quality, lack of information on current technological development, lack of managerial capacity and technical training and low capitalization and collateral security, among others. More often than not, they have to struggle to break even and remain viable economic entities.

Over many years, there have been several initiatives at the state and international levels aimed at stimulating growth within the manufacturing sector of some developing island states including those in the Caribbean, Africa and the Pacific. While it is necessary to have a framework for the development of some kind within which to provide support for small firms, there cannot be irrational duplication of structures and programmes similar to what exists in more developed countries, without having any relevance to the cultural and economic realities of small states.

Clustering / Networking in Small Enterprises

Several decades ago, two developmental approaches were popularised and promoted as the panacea to small and medium enterprise development viz clustering and business incubation. Although they have had varying results in developing states, their contribution to SME development needs to be rationally assessed to determine their effectiveness. The concept of clustering was popularised by Harvard Professor Michael Porter, while that of business incubators began in the US in 1959 and then spread across Europe and the rest of the world. Business incubators were developed to support start-ups and companies in the early stages. Clusters, on the other hand, were developed to support universities and research institutions.

Firms in a cluster produce similar or related goods or services and are supported by a range of dedicated institutions located in spatial proximity such as business associations or training and technical assistance providers. An industry cluster is different from the classical definition of industry sectors because it represents the entire value chain of a broadly defined industry from suppliers to end products, including supporting services and specialised infrastructure. Cluster industries are geographically concentrated and interconnected by the flow of goods and services, which is stronger than the flow linking them to the rest of the economy. Clusters include both high and low value-added employment, suppliers of raw materials, components and machinery.

There is no universal acclaimed definition of the concept of "clusters." The House of Lords Select Committee on Science and Technology, Third Report (1991) defined clustering as "a strong concentration of companies often in the same or related industries, sharing a common geographical location, and often in close proximity to important parts of the science base."

Clusters take place when there are economies of agglomeration (co-location). The creation of linkages among firms is one of the unique features of clustering. The benefits to be derived from clusters can emanate from a strong local customer base, infrastructure, local suppliers and the availability of skilled labour and risk capital. Clusters are also defined as networks of production of strongly interdependent firms (including specialised suppliers) linked to one another in a value-added production chain. They can also be considered as a critical mass of enterprises located in geographical proximity to one another. The practice of clustering takes a more realistic approach to the economy, in that it makes provision for market dynamics and modernisation. Conversely, the traditional sectoral approach primarily focused on firms of similar orientation.

The sectoral approach to clustering focuses on horizontal relationships and competitive interdependence (relations between direct competitors with similar activities and operating in the same product market). In the traditional sectoral approach, firms are hesitant to cooperate with their rivals, hence, they place emphasis on their direct and indirect competitors. This often drives them to constantly examine ways to diversify their product base so that they can forge ahead in the market. These firms are often preoccupied with lobbying the government for subsidies, protection and policies to limit rivalry and competition.

Additionally, clustering focuses on the importance of vertical relationships between dissimilar firms and symbiotic interdependence in the value chain. Innovations are generally stimulated by a horizontal struggle among competitors in the same product market. The role of vertical relations between suppliers, producers and customers in advancing relations is paramount. Other groups within the cluster include customers, suppliers, service providers and specialised institutions. Most participants are not direct competitors but share common needs and constraints. There is often wide scope for improvements in areas of common concern that will improve productivity and raise the level of competition.

Most developing countries owe their development to the private sector which consists of micro, small and medium enterprises that generate a large proportion of employment and income opportunities. However, what is noticeable is that their development potential remains untapped. As firms operate in isolation, they are locked in uncompetitive production patterns and are reluctant or unable to involve progressive business partners that could bring in new expertise and know-how. Clustering offers an opportunity to break this cycle. There should be no illusion that the existence of a cluster is a guarantee to the success of business enterprises. On close examination, many of the barriers to the growth of SMEs have more to do with isolation in non-performing clusters, which is manifested in unhealthy competition based on lowering wages, disregarding the welfare of workers, disregarding the importance of investing in technology and depleting natural resources. This form of competition results in the age-old problem of stagnation of clusters. Clusters also include support institutions, such as:

- business associations.
- business development service (BDS) providers
- financial service providers, including banks, development corporations, public authorities such as local, regional and national governments and regulatory agencies, training agencies including vocational schools, universities, etc.

The clustering of small enterprises has practical applicability for many developed countries where there are established industrial estates, technology/science parks and informatics parks that support research institutions. Firms that operate on these sites manufacture a range of goods from consumer products including high tech products to electronic components. For example, the Hsinchu Science Park in Taiwan, Detroit's auto industry concentration, computer chip production in California's Silicon Valley and Napa Valley's wine production. Generally, the sectoral approach to clustering is dominant. Clusters are associated with high rates of economic growth, technological dynamism, and a low rate of unemployment. There is a body of evidence that suggests that SMEs tend to innovate and perform better as clusters than as individual firms.

Small firms can cooperate by regrouping so that those with related activities can take advantage of common factors such as technology, skills and inputs, which will lead to improved productivity and enhanced competition. Firms rarely innovate in isolation, but in a network of production. Most innovative activities involve multiple actors and stem from combining complementary and specialised competencies and knowledge of various actors. For clusters to succeed in small developing states, they require the unswerving support of national institutions, including chambers of commerce, local governments, NGOs, producer associations, research and training institutions as well as development agencies.

Moreover, small firms in small developing countries need to develop a culture of cooperation built on mutual trust, build synergies and develop some level of symbiosis. By simply duplicating what exists in developed countries without understanding the nature of what drives relationships among firms, the interplay of the cultural, social and economic dynamics within the industrial environment and the ability to make the appropriate adjustment would result in failure.

It must be noted that size is one of the predominant factors that constrains the growth of SMEs. However, they are generally constrained by a plethora of other challenges that have stymied their growth for many years. Because of their small size, they cannot benefit from the economies of scale in the purchase of equipment, raw materials, consulting, etc., which would affect their purchasing power. SMEs have little influence on policies that will impact their development. Individually, they have limited bargaining power. They are also unable to take advantage of market opportunities that require large production quantities, regular supply of materials and compliance with international standards. They are unable to accept large orders for standardized products because they lack production and delivery capabilities. Due to resource constraints, SMEs are required to engage specialised support services at exorbitant costs. Notwithstanding the aforementioned constraints, the cluster environment offers tremendous promise for the growth and development of these small entities.

Where SME clusters exist, several characteristic features are common among them which serve to undermine the growth of these firms and their survival. Their production systems are often rife with antiquated technology and processes, which is a major barrier to productivity. The perennial challenge of firms lacking vital testing laboratories as well as knowledge about products and process standardisation is a major barrier for SMEs to produce products in compliance with international standards. Most of the enterprises in the SME clusters in developing states are producing products in isolation from other industries and markets because the relevant sectoral linkages are not established. Clusters of this type are prevalent in food processing, textile and garment, wood products and furniture, chemical industries and metal fabrication.

They have not developed the necessary backward linkages with producers and suppliers of raw materials and forward linkages with industry for further value-added. Knowledge of their customers and end-users of their products is often lacking and there is generally no diversity in the product range. There is also a tendency to produce similar products with the same quality. Because of the challenges and fierce struggle to preserve their narrow profit margins, small-scale entrepreneurs in developing countries are often locked in their routines and unable to introduce innovative improvements to their products and processes and look beyond the boundaries of their firms to capture new market opportunities.

According to a UNIDO report, "only a handful of SME clusters in the world are truly performing ones." Relatively few clusters or in the true sense of the word, quasi clusters in the small developing states have been able to achieve high and sustained growth rates. In many cases, they are trapped in a cycle of cut-throat competition, stagnation and poverty and are unable to spontaneously achieve the transition to innovation and growth. Consequently, appropriate policy support and assistance are often required. To arrest this deliberating trend and getting these SMEs on a path of growth would require appropriate public policy and support and other critical interventions. The question arises whether clusters can effectively survive in small developing economies given the complex nature of an organization including unique skillsets, specialised raw materials, components and dedicated suppliers that are required to make them work successfully.

The concept of clusters in small developing states does not have the same connotation as those in developed countries. What is generally referred to as clusters are not necessarily spontaneous developments. Only in rare circumstances are firms strategically located in specific geographical areas to create clusters. Because of land constraints, small firms are generally located in areas where there is available land. In some countries of the Caribbean for instance, there are historically designated areas that have been used as industrial estates or industrial parks where small enclave industries such as packaging plants, breweries and bottling plants are located. These firms bear no relationship with one another and survive on their own merit. The concept of clustering in small developing states differs remarkably from what obtains in more advanced economies of Asia and Europe. Considering the foregoing, the concept of clustering should be redefined within the context of small island states' socioeconomic and geopolitical reality.

Business Incubation; Challenges and Opportunities

An incubator is a physical facility with small work units providing a supportive environment for entrepreneurs and investors during the start-up stage or early stage of their business development. In other words, business incubation provides firms with intensive hands-on support to overcome the most common challenges that would lead to failure. They are designed to assist businesses to develop and grow to become viable and profitable entities and thereby increase their chances of surviving during the early stage of their development.

Technology incubators are a specific type of business incubator that focus on new companies with more advanced technologies. They are sometimes referred to as innovation centres, science parks, or technology centres. Technology incubators are not stand-alone ventures but tend to be affiliated in one form or another to public and private sources of research knowledge including universities, public research institutions, as well as technology-based firms. Within the framework of a technology incubator, information communication technology specific incubators have emerged. Although they are called by different names, their core functions and the services they offer mimic those of a general business incubator. The technology innovation centres as they are sometimes called support the growth and development of early-stage, technology driven, and technology enabled companies in the residential programme. Within the services sector, many firms have effectively used ICT as a tool to reduce costs, improve communications with customers, improve logistics, overcome distance and understand their competitive position. Firms within incubators should become self-sufficient as quickly as possible by focusing on revenue generating streams. In the meantime, the incubator should identify investors that raise start-up capital. The services provided by business incubators can be summarized as:

(i) Incubator space for tenants, laboratories, etc. at affordable rental on a flexible basis.
(ii) Shared services such as administration, shipping and computing.
(iii) Technical assistance and training in developing business plans, databases, marketing and access to start-up finance, often financed from an in-house fund.
(iv) Synergy through networking, entrepreneurs from physical proximity can benefit by exchanging experiences with one another as well as with the wider business community.

Business and technology incubators are primarily service-oriented. Technology incubators provide very specific and high value-added services. Large incubators can provide comprehensive services, but smaller ones provide referrals or links to other public and private agencies. In developing countries where there is limited venture capital financing, incubators can assist in building relationships that can provide much needed finance. There is no single model for a business or technology incubator. The objectives of these incubators often reflect the policy context in which they operate at the local, regional and international levels. The general objectives of incubators can be classified as follows:

(i) Economic development;
(ii) Technology Commercialisation;
(iii) Property venture/ real estate development;
(iv) Entrepreneurship.

Incubators offer several important services to fledgling enterprises in selected industries. They often rent premises and provide facilities to researchers, scientists and SMEs and lease land to technology knowledge-based companies. They also provide mentorship and coaching services, marketing and financial consultancy services along with organised workshops and business forums on business-related issues that are of interest to them. Specialised training is also being offered in the areas of project management, market research, strategic management and technology transfer facilitation. As discussed earlier, the main purpose of establishing incubators is to provide a nurturing environment for the success of businesses in their start-up stage. Incubators can be classified as follows:

(i) "Western Model"
(ii) An incubator "without a wall" providing technical services to tenants who do not occupy a shared facility.
(iii) Hybrid type Incubator providing both shared rental space as well as outreach services to existing enterprises in dispersed locations.
(iv) A "rural incubator" catering to the needs of non-metropolitan communities primarily in agro-processing related activities. In short, the Technology Incubator Centres (TIC) should identify and focus on local needs and resources.

A generic model of a full-service incubator offers the following services:

(i) **Incubator space**: offices or workshops, sometimes labs, available on a flexible, affordable and temporary basis.
(ii) **Common services**: including secretarial support, telephone answering, common reception, mailing facilities, access to computers (LAN) and other office equipment, meeting rooms and (in some cases) cafeteria facilities.
(iii) **Business counselling**, namely 'hands-on' assistance concerning business planning, training in management skills, access to accounting, legal services, marketing, financial expertise and ad hoc advice.
(iv) **Access to finance with specialist advice**: some operate their own seed and venture capital funds; access to specialist advice will be provided if an incubator does not have the skills and know-how itself.

The classical business incubator is an organization that has limited resources (people, money). Secondly, business incubators are constrained by their sponsors (government, or private investors) and act according to the goals for which they have been established. This happens if they are dependent on external financing and do not have independent income streams to be self-sustainable.

Incubators essentially can help to nurture new enterprises in their most vulnerable phases until they "graduate." That is when they are financially viable and freestanding to leave the incubator. They provide a range of services, from hands-on management and technical assistance and access to finance, to support services and infrastructure such as office space and communication facilities. Although their core competency is not focused on providing access to finance to SMEs, they can be a useful conduit for venture capital and establishing networks of investors.

Generally, there are two types of incubator models: "non-profit" and "for-profit" incubators. The former is typically established to promote economic development by creating the relevant infrastructure aimed at creating employment. These incubators generally sustain themselves through the rent and fee-based model charged to clients, which is highly subsidized and through complementary consulting and training fees. The second model usually involves taking equity stakes in the incubator's client companies in exchange for supplying office space and services. At the end of 2002, there were approximately 3000 business incubators worldwide, with approximately two-thirds based in North America and Europe and the rest around the world. In general, research suggests that the less an incubator relies on subsidies, the more successful and viable it would become. On average, every 50 jobs created by an incubator client generates another 25 jobs in the community.

It would be useful if some emphasis could be placed on rural incubators to facilitate disadvantaged groups such as women, youths and the physically challenged, but more importantly to arrest rural drift by providing employment and training in rural communities. Activities involving agro-processing, handicraft, furniture and woodwork, equipment and small engine repairs are well suited for such centres. These centres will facilitate the transfer of appropriate technology to relevant firms. In circumstances where a highly technological firm is established and would require specialized training, it could be housed within its confines. These centres will liaise with various technical, academic and research institutions that will provide information and input that is of importance to their clients.

The success of incubators is largely based on the objectives of the stakeholders that are sometimes not clearly defined. The mechanisms for measuring success do not necessarily relate to the objectives. Incubators are generally considered successful when they generate income for stakeholders, develop new businesses and move out, create jobs, diffuse technology and generate tax revenue. The role of networking cannot be overemphasized as an important element in the success of incubators. They may organise venture forums/fairs and bring together potential investors. Incubators cost money and take time, and it makes little sense if, for instance, 80% of the tenant firms are just surviving and are not growing, they can leave the incubator after 3 to 5 years. In other words, firms that should have been closed are kept alive on life support.

These initiatives are designed to make enterprises within the sector relevant and to reposition them for greater competitiveness in a changing world economy. In so doing, these firms would have a greater chance of competing successfully in the global market.

The concept of Business Incubation has gained popularity in some small states of the Caribbean in recent years. It is, however, not necessarily a panacea to avert the failure of start-up businesses. The model that is most commonly touted is one, where, within a designated facility, every conceivable business development service including training workshops, accounting, marketing, ICT services, counselling and advisory services are provided. Several reports have shown that most of these incubators in the Caribbean have to some extent failed. Where they have been deemed to be successful, they have become a burden to the state and some private sector resources. In most cases, they ceased to exist when funds from development agencies have dried up.

Government funded SME support agencies are rare in several small states. Those that exist tend to lack the strategic vision, resources, autonomy and linkages with successful firms to promote private sector development with an ICT advantage. Other assistance that is required by SMEs to help them to grow includes market information, legal and financial advice to foster entrepreneurship. To be successful, these firms need the following:

(i) A clear mandate and strategic vision,
(ii) In-house technical capabilities to execute that mandate,
(iii) Private sector involvement and the requisite personnel with entrepreneurial expertise to execute the strategy,
(iv) Performance metrics to assess an agency's successful interventions with the private sector.

It is perhaps timely to rethink the philosophy of Business Incubation in the context of developing states. It requires a model of business support in which the state in collaboration with the private sector will provide only those services critical to business start-up. The state institutions that are designated to offer services include the Bureaus of Standards, Small Enterprise Development Units (SEDU) and the National Development Foundations (NDF). This approach will reduce waste that is often associated with the provision of free services and will instead increase efficiency and help to develop a culture and mindset that will stimulate SMEs to work towards acquiring assets that they can use as collateral in the future.

Incubators have had high success rates in several Asian and European countries. They have contributed to growth within enterprises. For small island states, this concept could be customized to give rise to what could be called a "Business Support Programme." Within the framework of a Business Support Programme, several initiatives including clustering, benchmarking and the like, (where applicable and desirable) can emanate and thereby enhance the possibility of success for small firms.

In the current global scenario, small island developing states cannot take a monolithic approach to development. Those that lack natural resources must seek to broaden their capability so that they can respond to the changing world realities. Economic planning should take into consideration all the inherent disadvantages and limitations of a country's existential situation and devise innovative ways from which possibilities can emerge. The lessons of the Newly Industrialized Countries (NICs) of East Asia, as well as Switzerland, should be inspirational lessons for other small developing states. The foregoing issues present possibilities and options for industrial development by establishing the framework for small firms to be competitive. Initiatives such as Clustering and Business Incubation and their ancillary services are designed to create an environment where small and micro enterprises can grow and develop into successful enterprises, but only in circumstances that warrant such intervention.

The process of commercial innovation can be used effectively to revitalise ailing industries, particularly state-owned enterprises if the government is unwilling to privatise them. This approach will enable such enterprises to become more productive and competitive, thereby contributing in a more significant way to the social and economic development of the countries. Industrial development is a dynamic process and should encapsulate the major state resources and amalgamate them in an organised manner. This can be done through the process of technology transfer, notwithstanding that it has its merits as well as drawbacks. It should be carefully planned so that there are no adverse and debilitating effects on society. More importantly, whatever decisions are made, and resources expended regarding the transfer of technology should be able to realise maximum benefits to any country.

Incubators on the other hand are often deficient in setting clear and unfailing performance metrics and success criteria for companies to "graduate," and they often struggle to find ways to become financially sustainable and less reliant on subsidized funding.

Capacity Building within Private Sector Organizations

In many small states, the public sector is the main employer of labour because the private sector is generally underdeveloped and cannot provide employment for a substantial proportion of the workforce. This situation has political implications for the poor and vulnerable groups. To address this vexing question of unemployment, the government is obliged to provide funding to support small business ventures on two fronts (i) either to arrest the decline of those on the verge of collapse by stimulating some level of growth or (ii) to assist would-be entrepreneurs in their quest to start new business enterprises. Most of these programmes are not well thought out and rationalised and are designed with no long-term strategy in place.

For the most part, the relevant benchmarks are not established, and lessons learnt from similar programmes in countries with similar social, economic and geographic characteristics are not adopted. In such situations many of these programmes are doomed to failure, resulting in a waste of public funds. To survive the vagaries of the current global market, small enterprises need the relevant support infrastructure that is driven by a philosophy of change. This approach should be responsive to the challenges of global competitiveness and driven by innovation in technology, value-added initiatives, product design and marketing.

Entrepreneurial development requires access to financing, information, and various types of support services to create and operate viable business enterprises. Enterprise managers need similar access to operate their businesses efficiently and competitively. The existence of institutions such as universities, consulting firms, accreditation and standardisation bodies and vocational training institutions can play major roles in providing support services for the private sector growth and competitiveness.

It must be emphasised that well established private sector entities have the relevant resources and in-house capacity at their disposal that they have amassed over time to provide the business support services that are fundamentally necessary to enable them to become competitive entities. Their survival is not dependent on support from the state or development agencies. On the other hand, the situation regarding SMEs is far more challenging. They require state and business support intervention at various stages of their development. The productivity gap between medium-sized manufacturing enterprises (MSMEs) and larger enterprises is wide. On average, SMEs productivity in Latin America and the Caribbean is about 40% lower than that of larger enterprises. Improving productivity in SMEs is dependent not only on access to credit but also on the availability of a diverse set of non-financial services, essential to the development of effective and efficient business practices and production processes.

One of the main goals of business support services is to make support services readily available to micro and small enterprises. Such services encompass a range of strategic linkages, cooperation, and partnerships aimed at strengthening symbiotic arrangements among small firms. These arrangements should serve to embolden and strengthen the capacity of small firms to develop their productive capacity through specialised technical assistance in the application of new technology transfer mechanisms, operationalizing production management and organization techniques and methodology for leveraging financing.

For the most part, small firms need assistance in building relevant entrepreneurial capacity as well as specialised training in human resources appropriate to identified needs. Among the most vulnerable small enterprise groups that have struggled over the years in these areas are agribusiness, food, textiles and garment-making, information and communication technologies, construction and handicraft.

Rather than focusing on the notion of "cluster," the emphasis should be placed on establishing networks among small firms that are already in existence, although they are not necessarily located in the same locality. What is important is to link these small enterprises through networking, where they can derive several benefits. Through networking and institutional support, individual SMEs can address challenges relating to their size and thereby improve their competitive position because they are in the best position to assist one another to initiate and develop programmes to enhance their wellbeing. Effective relationships can be developed among independent entrepreneurs based on collaboration and production integration even when the entrepreneurs have no previous knowledge of one another.

The case of Jamaica and Honduras are classic examples where specialised institutions were established to act as networking promotion agencies. The programme involves two main features (i) institutional capacity building and (ii) network promotion. The programme supported the SME sector by providing critical services to the sector based on a few identifiable needs. Some of the projects initiated include joint purchasing of raw materials, joint establishment of retail shops for finished products launching new production lines, product or process specialization, sharing of large orders (including public procurement), and the creation of new enterprises which complement existing production facilities.

E-Commerce; A Tool for Competitiveness

Electronic commerce, commonly called e-commerce is one of many tools in information communication technology that can be used to facilitate business transactions across frontiers. The concept of e-commerce is about using the Internet to do business faster and better. Web Commerce is a payment gateway that provides the interface between one's website, the credit card processing company and a merchant account/bank.

In February 2019, 76 members of the WTO expressed their intention to commence negotiation on the adoption and use of e-commerce as a tool for further development among its member states. To date, few small states have joined these negotiations, although they are eligible because they are WTO members. Many of these states continue to grapple with economic development. The tourism sector in most of these small developing states declined to an all-time low as a result of the impact of the COVID-19 pandemic. However, in recent years, there has been a resurgence in the growth of the sector.

The manufacturing sector, which is comprised of micro, small and medium-sized enterprises, has perennially been experiencing several challenges that have stymied its growth as a major foreign exchange earner. Among the major challenges facing the sector are inadequate supportive policies, technology diffusion, regulations, and access to markets for its products. For the most part, market access is primarily focused on local and regional markets. It is within this context that e-commerce offers the potential as a potent tool in digital trade in goods and services. Such trade can take place within the confines of marketing, distribution, sales and delivery of goods and services through electronic and digital means. The potential of e-commerce as a catalyst for economic transformation in small developing states cannot be overemphasized. It is therefore critical that these states explore the possibilities of adopting e-commerce for online transactions.

For small states to transition from a knowledge-based economy will require serious planning and government support regarding infrastructure resources and appropriate competencies. Countries need to decide how they will use their comparative advantages to create opportunities and close the gap regarding skills, knowledge, capital and technology. In recent years, the focus of many small developing states particularly those in Asia has shifted from manufacturing to information technology-oriented industries such as call centres and outsourced business processing operations that offer better paying jobs and a greater infusion of ICT technology. There are spillover effects to be gained by giving long-term commitment to foreign businesses, which will result in greater levels of interaction with the local workforce through specialised training.

The lack of digitization of processes and systems in many developing economies has increased their vulnerability to external shocks, limiting continuity in the operations of the public and private sectors and the inability of individuals to conduct daily transactions in a manner that would limit contact with one another.

The rise of global digital technologies and the emergence of digital economies offer a unique opportunity for small developing countries to accelerate economic growth, provide employment opportunities, enhance public service delivery and build resilience. Developing countries can chart a new path by working collaboratively to create a more deeply integrated and dynamic digital economy. These new technologies can enable countries to empower their citizens with soft skills through which they can find meaningful employment in a knowledge and services driven global economy. Businesses and institutions would have the opportunity to build a future in which seamless and efficient public services are available at the touch of a screen from even the remotest island. This new paradigm of business venture in which businesses and entrepreneurs are pushing the frontiers of innovation, creating new jobs, and accelerating a region's economic growth can have a profound impact on the economy of developing states.

Cross-border e-commerce electronic transactions are made among users from different countries, enabled by digital platforms. In 2015, global e-commerce (internal and cross-border) trade reached an astonishing US$ 25.3 trillion. The evidence shows that the consumption of products imported via online platforms has been growing exponentially, unlike the export of goods using other modalities which have not increased to the same extent. This situation suggests that there is a gap between what is imported and what is exported. The challenge for small businesses is to invest in the requisite infrastructure to enable them to utilise the online platform for business transactions.

In practical terms, several measures will have to be implemented by many small states to establish a framework that will facilitate a viable and authentic cross-border e-commerce platform. Firstly, an e-commerce platform will require a digital ecosystem including broadband connectivity, the availability of information and communication technologies (ICT) and appropriate technical skills. Secondly, enabling e-commerce legislation preferably on a regional level. Thirdly, appropriate efficient logistics and customs procedures to facilitate cross border e-commerce in goods. In most small states, custom procedures are outdated and need to be modernised and digitized through paperless trade and the creation of a trustworthy online vendor programme. Fourthly, the system of payment needs to be secured and strengthened to foster confidence among its operatives. This area poses few problems since credit and debit cards are in vogue for online payment. Despite the use of these cards, there may be challenges in using bank cards.

E-commerce is often regarded as hi-tech and there is an apparent inhibition on the part of some persons and organizations using it for business purposes. There are several applications for which e-commerce technology can be used effectively by small and medium-sized enterprises for transactional business purposes, notwithstanding the sophistication of the transactions.

E-Commerce Infrastructure

Some developing countries have made commendable progress in developing their digital foundations, most notably in upgrading digital infrastructure. However, a lot more remains to be done to build the capacity to make e-commerce fully functional. As a case in point, the World Bank has provided financing to build a new fibre optic network across several small states of the Caribbean. The network is now capable of delivering high-speed broadband services to government offices, schools and other public services facilities, thereby creating the opportunity to modernise and digitize government internal operations and to offer more efficient and convenient digitally enabled services to citizens.

Moreover, the infrastructure is being utilised by one of the services providers to offer mobile and high-speed internet services to private individuals and businesses. This new initiative would result in the creation of opportunities that will transform the way business is conducted and how individuals access information, services and employment opportunities. Further efforts are now needed to capitalize on these opportunities.

While recent fibre-optic infrastructure upgrades have improved the capacity and reach of networks, access to broadband services remains challenging for many individuals and businesses and service quality remains unreliable. The cost of a basic mobile broadband plan is above the average of peer countries at similar levels of per capita income. Perceived value for money is still a farfetched goal, because of the low quality of broadband services and limited digital content. These challenges disproportionately impact the poorest individuals, micro and small enterprises and vulnerable or marginalised groups, which continue to be locked out of the digital economy as a result.

The liberalisation of the telecommunications market in many small island economies is necessary and desirable to eliminate any existing monopoly and encourage more service providers to participate in the provision of services. This decision is geared to improve the quality of service and reduce the costs of these services to the public. Liberalising the telecommunication market in many developing countries is not without contention. However, it is imperative that governments enact the requisite legislation to support the necessary policy decisions to liberate the telecommunication market. It must be noted that there was universal unanimity among small states to liberalise and deregulate the telecommunication market from the exploits of dominant services providers.

Small developing states that are on the threshold of fully embracing information and telecommunication technologies must be prepared to address some serious human resource drawbacks that cannot be ignored and dealt with naively. The issue of establishing ICT related business should be approached with a level of sensitivity that will negate imprudence and squander. The lack of experience in selected business ventures should be compensated for by acquiring skills and experience externally (if that were the most logical option) supported by the appropriate policy framework and other modalities that will guide successful operations. It is within this context that government should seek the guidance of experienced and trained practitioners in devising a programme of activities that will forge e-commerce forward.

Constraints within the E-Commerce Sector

Progress regarding the active involvement of the private sector in the promotion of e-commerce is often hindered by factors outside the domain of the government. E-commerce is often perceived as being demand-driven by financial institutions such as banks, building societies and cooperatives that should act as conduits through which transactions can be facilitated.

These factors may engender enormous capital costs, particularly if the government should assume the responsibility of devising programmes and strategies that will give impetus to the growth of a viable e-commerce sector. Some governments have responded favourably by enacting appropriate legislation like the Electronic Transactions Act to address issues such as the legal requirements for electronic transactions, accreditation, consumer protection, computer related crimes and other procedural matters.

Despite its tremendous advantages, the growth of e-commerce in some jurisdictions has been seriously retarded because of the lack of awareness, training, appropriate policy to address security and other critical issues within the sector, software requirements and all the other requisite tools that will facilitate the growth of the subsector. The sector is further constrained by the lack of a sufficiently large clientele that will make e-transactions profitable and a general lack of confidence in the system due in part to the aforementioned factors. This thinking explains the lackadaisical approach of some institutions to fully embrace e-commerce as a major development tool. On the other hand, even manufacturing enterprises hesitate to use e-commerce as a tool, which will enable them to reach a wider market with their products.

The cost of website creation and operating services (creating, maintaining and hosting a website) could be a deterrent for small and medium-sized enterprises, particularly if the long-term benefits were not well understood. In this context, there is a need to establish the relevant infrastructure to facilitate training and thereby build the relevant capacity so that the cost of these services can be minimized. Online portals provide an operating framework that will enable business entities to interface and transact with customers, suppliers and employees. The graphics, content, security, and business intelligence will be tailored to the end users and sent via the most appropriate method. In other words, there are no sophisticated system requirements to develop an e-commerce system successfully. All the supporting institutions and ancillary service providers are market ready. These opportunities and benefits have a potential trickle-down effect on wealth creation at the individual and national levels. It is for this reason that the promotion of e-commerce should be given momentum so that opportunities can be inexorably pursued within the different sectors of the economy.

The Role of E-commerce in Modern Business

The importance of E-commerce in a global retail market is undoubtedly a desideratum for all developing states, not least the small island states. Since the advent of the internet, there has been a phenomenal increase in the use of online transactions. Such transactions have taken the form of Business to Business (B2B), Business-to-Customer (B2C) and Business-to-Government. Shoppers now have the opportunity to take their shopping carts to virtual shopping areas to purchase goods that hitherto were not available within their own jurisdiction. The number of online shoppers has increased exponentially over the last couple of decades due to the availability of internet supporting devices as well as access to the internet.

Cross-border E-commerce is a new driver of export trade for SMEs, particularly where regional exports have seen little growth in terms of volume and have been concentrated on a small number of products. The availability of computers and access to the internet has drastically reduced international transaction costs by bringing buyers and sellers closer together and making foreign markets accessible that were hitherto unattainable.

E-commerce is among the many tools in information communication technology that can be used to facilitate business transactions across frontiers faster and more efficaciously. Cross-border e-commerce has the potential to boost exports and promote diversity in goods, since companies that buy and sell online tend to export more than those that do not. They also sell their products and services to more markets. Additionally, these companies tend to survive longer and have higher productivity rates and wages. Consumers also benefit from e-commerce because of the time they save on purchases and the fact that they can access a wider range of products and services at lower prices.

There are multiple benefits to be derived from e-commerce that small states should not continue to ignore. With increased competitiveness in the world market, business entities with both product and service orientation should as a matter of necessity embrace E-commerce as an integral tool in their procurement activities. Such activities could include acquiring spare parts, supplies and equipment as well as selling their products electronically. Fundamentally, enterprises establish a web presence in the form of websites that will increase the effectiveness of marketing and advertising products on the World Wide Web.

There are increasing market access opportunities for micro, small and medium-sized enterprises as well as lucrative opportunities for individual entrepreneurs. Among the benefits to be derived from viable e-commerce trade are:

(i) Lower costs, cheaper overheads and higher profit margins;
(ii) The ability to market online with links directly to the store of one's choice;
(iii) Stores remain open 24/7, allowing for increased trade;
(iv) The global reach of the internet enables one to access new markets;
(v) Improving logistics;
(vi) Widening consumer choice.

It is inconceivable to imagine that small states that are economically disadvantaged will forgo such enormous opportunities by not taking their business to the online space. Digital technology allows SMEs to reach global markets and access knowledge networks at a relatively low cost.

Through the digital transition, SMEs can be afforded new opportunities to enhance their competitiveness. Evidence has shown that small enterprises have not been able to capitalize on the opportunities that digital technology has afforded, because they have been lagging behind in adopting these technologies.

It can, however, be argued that the cost of website creation and operation services including creating, maintaining and hosting a website could be a deterrent for small and medium-sized enterprises, particularly if the long-term benefits are not well understood. It is within this context that the government should provide leadership in assisting to build the appropriate capacity so that the cost of these services can be minimized.

Although many enterprises in developing states have a web presence, they, however, lack the ability or the interest to adapt to sophisticated applications including the e-commerce technology and Enterprise Resource Planning (ERP) application that will enable them to manage information flows. Perhaps the lack of resources to invest in these technologies may be the underlining reason. The ICT revolution has now made it possible for small firms to develop collaborative networks with other businesses and research institutions to enhance their innovation prospects.

Information technology cannot flourish without relevant institutional strengthening because of the highly competitive nature of the sector. Hence, it is in the interest of any country to develop its human resource potential with the appropriate skills that will maximize the available opportunities. The government should also demonstrate commitment by providing the infrastructure and policy framework for the continued growth of the sector.

In many countries, this philosophy is manifested through capacity building within the education system by effecting changes in the curriculum to reflect the needs of the growing sector. The establishment of appropriate technological institutions is a desideratum designed to provide the relevant training and thereby build capacity for emerging and existing industries. Consequently, it is incumbent on all governments to ensure that there is a cadre of appropriately trained personnel that will form the nucleus of the workforce for the various opportunities within the sub-sector.

Fundamentally, the development of a strong and viable ICT sector will become a catalyst to promote entrepreneurship within various information technology sub-sectors and promote opportunities for job creation. All efforts should be made to maximise the benefits of the available technology by building an efficient and reliable internal telecommunication and information system to provide quick response time to prospective entrepreneurs and the general needs of the citizenry. Such a system should be designed in a manner that will ensure there is a reduction in waste in terms of paper, time and personnel. An effective functioning ICT system should facilitate easy access to information promptly and thereby facilitate services that would otherwise have taken an inordinate length of time.

The rapid growth in the ICT sector worldwide and the success of countries such as Singapore, Taiwan, and Hong Kong in using ICT to drive development have attracted several countries seeking to replicate that success. Many small states have attempted to use the ICT sector as an engine of growth but with limited success. This is primarily due to the myopic strategic approach to entering the sector at the lower end of the market in areas such as call centres and telemarketing. These activities depend on low wages and low-cost connectivity, neither of which some jurisdictions could supply. Attempts to move up the value chain also floundered on poor telecommunications infrastructure and inadequate human resources.

ICT, A Business Facilitating Tool

The widespread use and application of information communication technology has allowed many small developing states to use related technologies as a tool in the quest to diversify their economic activities. Against the backdrop of the current fragile industrial climate and the partial demise of some traditional activities, there is a plethora of challenges to which countries are required to respond so as to maximize the use of ICT in their economic diversification programmes. It must also be recognised that the fast growing information telecommunication sector has created a new set of challenges and opportunities particularly for the manufacturing sector. The survival of the manufacturing sector will, to a large extent, be dependent on the industry's ability to integrate ICT into its manufacturing functions to maximize its resources and achieve optimal efficiency in production.

Governments in most small developing countries recognise the tremendous benefits that can be derived from the use of ICT if the sector were rationalised and streamlined in a manner that will allow it to develop and flourish competitively and thereby attract foreign investment. The sector requires a radical approach to establish the requisite framework within which a vibrant information communication technology industry can strive. With this imperative in mind, many governments have taken steps to liberalise the telecommunication sector through strategic decisions resulting in (i) the liberalisation of the telecommunications market, thereby encouraging the participation of other major service providers (ii) the creation of new entities to strengthen existing institutions to build relevant capacity to facilitate new development in the sector (iii) development of policy guidelines to give currency to activities that will forge the sector forward.

It must be emphasised that no quick and sustainable outcome will emanate from isolated investments in information communication technology. Although some jobs will be created from investment in information communication technology in the short term, realistically, it cannot be anticipated that the jobs created will have a significant impact on the employment situation in some states. There is a preponderance of evidence that shows that most of the jobs created including employment at call centres, data input, retrieval and sorting and bibliographic coding are basically at the lower level of the technology spectrum.

Creating opportunities for high end ICT jobs will require higher levels of investment targeting higher levels of professional skills in the areas of computer programmers, system analysts and designers, information system managers, technicians and maintenance engineers. These skilled professionals will undoubtedly provide the requisite support to a progressive and sustainable information technology programme which will have enormous scope for further employment.

As more small business enterprises turn to ICT to support their day-to-day operations, legislation should be kept attuned to improve its use. As more legislation is harmonized within a regional framework compatible with international standards, the greater the benefit to these economies and their attractiveness to foreign investors. The legal enforcement of electronic documents, protection of intellectual property and privacy, further liberalisation of the telecommunication market, development of a framework to support electronic transactions and enforcement of electronic documents and contracts are among the priorities in this sector. Laws on intellectual property regarding the Internet-in code, data, music, or other contents have important implications for businesses selling in a multi-jurisdictional market. Keeping abreast of these changes and providing up to date enforceable legislation will improve the ability of small states to transact with the rest of the world and attract foreign businesses. Mauritius for example, has been a leader in this area of development in Africa and was able to convince Microsoft, Hewlett Packard, and IBM to establish their regional headquarters on the island.

In developing a global supply chain, a major barrier is communicating across cultures and different time zones. This hurdle has been overcome and facilitated by modern ICT. The availability and accessibility of modern telecommunication technologies has made it considerably easier to integrate the supply chain across several regions of the world. ICT including satellite, cellular devices and the internet has made global communication available and continued development in wireless technology has made these services cheaper and more affordable. The advent of the Internet has made it possible to create real time linkages among global firms in which they can link their logistics and production and target potential suppliers and customers. Firms around the world are now able to link their production scheduling, inventory control, and logistics through the use of tools such as Enterprise Resource Planning (ERP and Customer Relationship Management (CRM) software. Barcode technology has also been used effectively for product scanning and tracking of materials in the supply chain.

Chapter 7

World-Class Manufacturing & Management Techniques

It is well known that fierce competition often results in positive outcomes but can also have some negative drawbacks. Quality products and services, efficiency in production, productivity improvement and profit maximization are benefits that are directly related to a healthy competitive environment. On the other hand, rapid changes in product demand, regulatory stipulations imposed by mega trading blocs and high cost of capital, could inhibit improvements in product technology and render some enterprises uncompetitive. The result of the latter set of circumstances could lead to bankruptcies, insolvencies, mergers and acquisitions as well as other actions that characterize present-day SMEs in the business environment.

The fierce competition which exists in the marketplace today requires many firms to undertake detailed planning of their investment activities with shrewd allocation of resources. The economic and business environment is not conducive to haphazard and inefficient planning. The days of fortuitous approach to decision making and management are well into the past. The nature of the business environment has become more innovative, competitive and complex.

Most organizations regard sound corporate goals and strategic objectives outlined in their medium and long-term plans as fundamental imperatives to their business success. Profit maximization is pursued with the most thorough and rigorous management policies and decision-making techniques such as discount factor methods, return on capital employed and management simulation techniques, among others. Even small charitable contributions should also be planned and accounted for to justify the use of funds.

In a dynamic and robust business environment, changes in product demand, price, corporate strategies, regulatory stipulations and economic pressures often combine to engender the number of bankruptcies, closures, mergers and acquisitions, which seem to characterize the environment of present-day business organizations. As a countermeasure, flexibility and differentiation are often built into product planning and design as a competitive market strategy. Without a proper investment strategy, a high possibility exists that a firm will languish and eventually go out of business. Furthermore, a ubiquitous recession that so often strangles many small developing states seems to suggest that only those firms which have the technical and managerial capability are highly likely to survive in the market.

It is against that background that new capital investment must be carefully planned, thereby incorporating factors such as risk and uncertainty, as well as changes in various monetary factors including interest rates, the time factor, inflation and corporation tax that may influence the success or failure of an enterprise. The entire modern business landscape has become so revolutionised that by necessity it should embolden a technological management approach to planning and decision making. It is this approach that enables a manager to effectively combine the resources available using proven technology to produce a product at the least cost to the organization.

The Concept of World-Class Manufacturing

Over the last couple of decades, the term "World-Class" has become a buzzword in manufacturing and business circles. Although different people may give various connotations to the term, an appropriate definition should focus on excellence within the entire gamut of manufacturing. "World-Class" is described as "being the best in your field in the world." Characteristically, the word "best" in this context relates to the following conditions:

i. Product design and performance;
ii. Quality and reliability;
iii. Shorter lead times and more reliable delivery performance;
iv. Least manufacturing cost (this is where one competitor can undercut the other on price or spend more than they can afford on research and development or advertising);
v. Customer service performance makes customers bend backwards to buy from one competitor rather than the other.

A world-class firm or corporation connotes the idea of having the ability to adapt quickly and efficiently to change. These are essential features that form part of the culture and reflexive behaviour that allows an organization to compete successfully. Some of the key features in defining a world-class organization include quality, customer responsiveness, design leadership, market share, return on capital employed, cost leadership, growth, sales per employee and earnings per share. An organization needs to place emphasis on these characteristics through its goals and policy design, which will become the building block for a successful organization.

The concept of manufacturing is an economic term, which implies creating value by applying useful mental and physical labour in the production of goods and services. Production processes are combined to form a manufacturing system. The manufacturing system is a collection or an arrangement of operations or processes that are used to make a desired product or component. This is inclusive of the equipment used in the various processes. The production system, which includes the manufacturing system and services, is the highest in the hierarchy. The production system involves people, money, equipment, materials, supplies, markets, management and the manufacturing system.

There are numerous techniques and philosophies that are available and can be adapted to enable firms to achieve world-class status. The following techniques have been identified as the driving force behind world-class manufacturing. They include:

- Total Quality Management (TQM
- Just-in-Time
- Lean Production
- Teamwork (Total Employee Involvement
- Continuous Improvement

Other techniques have also been used to good effect to overcome specific problems that can be regarded as barriers to achieving world-class status. Among them are:

- Manufacturing Resource Planning (MRP11)
- Computer Integrated Manufacturing (CIM)
- Total Productive Maintenance (TPM
- Automation and Flexible Manufacturing Systems (FMS)
- Quality Accreditation (BS5750, ISO 9000)
- Simulation
- Design for Manufacture

The Industrial Revolution, which triggered the development that transformed Great Britain between 1750 and 1830 and consequently other developing countries, has undoubtedly continued to influence revolutions in industry, notably manufacturing. Several innovative initiatives have taken place in manufacturing and management over many decades that have radically transformed production and management processes.

In the 20th century, Henry Ford revolutionized the automobile industry by introducing a system of coordinated assembly line operations. Ford's success led to the widespread adoption of the mass production system in manufacturing. This system became a pioneering ingenuity that triggered a plethora of manufacturing techniques and philosophies that are instrumental in the new paradigm of production, manufacturing and management. Some of the techniques and concepts that will be discussed have been proven to be effective over time and have produced desired results as evidenced by performance enhancement and increased efficiency and production. Numerous companies in Europe, Asia and the United States have attained world-class status for some time now. It was not until recently that Japan and other East Asian states have been setting new standards in manufacturing through management techniques and philosophies.

These new techniques have also been used widely by many companies with tremendous success. Some of the changes that have taken place in manufacturing in recent years include:

- Worldwide competition that is driven by first-class companies
- Advanced manufacturing process technology
- New manufacturing systems structure, strategies and management.

The secret of success in manufacturing is to build a company that can respond to the changing needs of customers and deliver products of superior quality promptly and at the least possible cost. This needs to be done with a level of flexibility that enables the company to respond to changing demand. It must be established that the challenges in manufacturing and management are numerous, complex and diverse.

As technology changes, so too are the challenges associated with it. Hence, there should always be new approaches in the production of goods and services to respond to these changes in a cost-efficient manner. The list of modern manufacturing/management techniques is not exhaustive but has been increasing steadily as new studies and methods reveal a greater level of success in motivating employees, dealing with customers as well as new mechanisms to respond to market demands.

Most of the methods and techniques practised by industry today are fashioned by the corporate objectives of enterprises. Furthermore, technologists and management scientists are constantly exploring ways of improving productivity by reducing waste, minimizing set-up time, motivating employees, establishing highly trained workforces, making the job safer, tracking material flow, using production planning, increasing efficiency and improving quality. Notwithstanding nomenclature, some techniques bear similarities while others are distinctly different. Many manufacturing enterprises that have achieved world-class status have developed and incorporated these systems over time, while others, particularly from developing countries, have transferred the technology in its embodied or disembodied form. Such technology should be made appropriate to their specific needs and circumstances.

It must be noted as a matter of caution that implementing a particular set of techniques will not necessarily make a company world-class. Several considerations such as production capacity, vis á vis market demand, resource availability and support mechanisms (such as clusters of suppliers' firms) should be assessed regarding the strengths and weaknesses of the organization. On the other hand, the question of size may not necessarily preclude a firm from becoming world-class. For example, small vineyards in France are world-class wine producers because they have adopted appropriate manufacturing and management systems.

This chapter will review selected manufacturing and management techniques and concepts to determine their applicability to manufacturing enterprises in small developing countries. It is intended that this exercise will, firstly, inform the selection process (by firms) of the importance of focusing on the functionality of a particular technique within a given culture and socio-economic environment as well as the production capacity relative to corporate objectives and demand.

Secondly, the benefits of the techniques will be discussed to determine whether they are realistically achievable within a given production environment. Thirdly, other techniques will be discussed on their merits so that information concerning other choices will be available.

It is expected that manufacturing enterprises will not make radical choices that will lead to unanticipated changes in production. In the long run, benefits relating to corporate goals and objectives will be realized as decisions regarding the choice of techniques are systematically and judiciously examined.

Evolution of World-Class Manufacturing Techniques

In the years preceding the emergence of world-class manufacturing, attempts were made to encourage companies to achieve international standards of excellence in quality by awarding prestigious prizes including the Malcolm Baldridge Quality Award (U.S), Deming Prize (Japan), ISO 9000 and the European Quality Award. The main criteria varied based on the focus of the award. For instance, the Malcolm Baldridge Award designated 250 out of a possible 1000 points for customer focus and satisfaction. In the case of the European Quality Award, 20% of the points were given for consumer satisfaction and the Baldridge Award on the other hand, stressed quality. Although not all winners of these awards have been competitively successful, they nevertheless should be regarded as helpful in encouraging the achievement of strategic manufacturing success.

With the advent of total quality control and the Toyota system, Japan rose to industrial prominence between the late 1960s and early 1980s. During that period, it was reported that many U.S.A businesses were experiencing problems with productivity, evidenced by low inventory turnovers in major companies like Kodak, IBM, and Ford. This was primarily because they were not customer focused.

Innovations in management are said to have occurred in three distinct decades. The 1970s, as Japan's decade, was marked by innovations such as total quality control, Just-in-Time, Kanban, total preventive maintenance, quality function deployment and employee involvement.

America's decade, which spans 1985 through 1995, was associated with benchmarking, reengineering, design for manufacture and assembly and point of sale technology. Within a decade, Japanese best practices in product development spread rapidly into Western firms. During the decade of the 1900s, the emphasis had been on the scientific management revolution. As a result, there has been a multiplicity of innovations in manufacturing management from various parts of the world.

Features of World Class Manufacturing

Until recently, manufacturing management had not measured up to world-class standards. It was believed that critical management functions and responsibilities reside with a few top managers and that information dissemination was vertical (top-down). For world-class management to be effective, the organization should build versatility, adaptability and flexibility within its ranks. Every employee must be involved in the management of the enterprise and should be committed to continual improvement. This new approach has effectively removed feelings of indispensability from management that may have been caused by a few persons in the organization who were highly skilled and influential and who enjoyed the status of belonging to the top tier of the management hierarchy.

A world-class manufacturing approach would avert any crisis in management if a few personnel in the upper echelon of the organization had to leave. Some organizations are insecure in their ability to cope with adverse market situations and risks, including loss of market share, failed product lines, changes in technology and new product substitutes. These factors could render products obsolete and a multiplicity of other circumstances could occur that could create uncertainty and loss of confidence in an organization. The ability of an enterprise to respond quickly to challenges in the market using world-class manufacturing tools is vitally important because they provide options to address specific problems related to manufacturing. Many western manufacturers have adopted world-class manufacturing formulae and have achieved satisfactory results similar to those that the Japanese have achieved earlier. For instance, problems related to long lead time and high defect rate can be addressed using a Just-in-Time system.

The importance of achieving world-class manufacturing status or a satisfactorily comparable level cannot be overemphasized. Small manufacturing enterprises like those that are common in many small island states are struggling to grapple with the current harsh economic realities as well as the dynamic global market scenarios that present enormous challenges for the survival of these enterprises. Many firms with extremely low capacity utilisation and poor and archaic manufacturing techniques have been able to survive in the past as a result of the import substitution policies of some governments. These policies have placed restrictions that confer on the manufacturers a monopoly position in the domestic market. Nonetheless, the principal dilemma confronting the government in shaping its import substitution policy in this sector stems from the poor performance of most of the domestic-oriented firms, in terms of capital efficiency and cost competitiveness.

World-Class Information Systems

Information technology is an important cross-cutting tool in every aspect of development. In the production of goods and services, companies of various sizes have achieved a competitive edge by implementing appropriate manufacturing information systems. New products have been brought to the market in record time due to the use of advanced computer technology. Information technology, when used effectively as a tool, can give an organization an advantageous competitive edge in the market. In a modern-day competitive environment, every organization needs to collect, process and store data. These routine and onerous tasks were once done manually and required considerable time and paperwork. Today, companies use automated information systems to perform key functions. Information systems are often used to support internal departments in a wide range of applications including sales and marketing, logistics, customers and suppliers' interface, among others. Businesses can benefit from information systems by saving time and money while making faster and smarter decisions.

Other benefits from these systems include improvement in customer service, reduction in lead times, fewer stock-outs and missed deliveries. Just-in-time (JIT), total quality management and flexible manufacturing systems are said to have the greatest impact on information requirements.

Information technology accounts for a significant proportion of industrial costs. Information gathering, storage and maintenance can be extremely costly. The value of information lies in how effective it is as a decision-making tool. Consequently, there are additional costs in transforming raw data into usable management information. The use of computers in industry can be considered a risk or an opportunity. When they are misapplied, computers become a high-cost factor for any company, they also incur large overhead costs and the level of inefficiency increases. Such misapplication defies the whole objective of maintaining an information system. In many organizations, computers extend down to the shop floor where various components of production equipment such as robots, material handling devices and sensors are embedded with computers or electronically related equipment.

World Class Manufacturing Strategy

World-class manufacturing requires a unifying strategy to meet the challenge of current and future market demands and thereby stay in business and retain competitive advantage. Apart from understanding customers' needs, an enterprise should develop internal mechanisms to respond to the changes dictated by what customers want. It requires a paradigm shift in everything the firm does. In other words, an organization should be able to bring all the different facets of its operations (marketing, production, design, purchasing, etc.) together in an integrated way to achieve world-class aspirations and other organizational goals.

Before embarking on any costly investment involving sophisticated technology, management must first decide what the firm is going to produce and to whom it is going to sell. There must be a general commitment at every level within the organization to be the best in the world. This can be achieved by developing an effective management strategy and incorporating the appropriate technology and techniques that will be required to make it world-class and a market leader. The strategy should be elucidated and understood by all. Ambiguity and doubt in setting priorities could give rise to conflict, particularly among top management. Specialised training programmes should be designed for those who will be involved in key strategic projects.

An organization should meticulously develop a manufacturing strategy that will take cognisance of long-term changes in the market that could potentially affect the product as well as the organization. In so doing, the firm must understand the market in which it wants to operate. An understanding of the market will enable it to decide on the most appropriate marketing strategy to pursue. It must then decide on what world-class manufacturing techniques will provide the efficiency and cost-effectiveness that are required to be successful.

A world-class approach invariably means putting the needs of the customer first. An effective strategy will therefore address the question of how the organization will satisfy those needs at the least cost. Some flexibility should be built into the strategy to allow the organization to respond to changes in the market in an alacritous manner. There are several dimensions of customer satisfaction that include fitness for use, reliability (the life aspect of quality), value for money, after-sales service and support, packaging, customer information and training, maintainability, variety and speed of service.

As was mentioned earlier, numerous techniques and philosophies are available to enable companies to achieve world-class status. Rather than following a general arbitrary approach in discussing the characteristics and peculiarities of the above world-class manufacturing techniques, they will be discussed under three broad headings viz. (i) World-class management techniques and philosophies (ii) production planning and control and (iii) quality management and assurance techniques.

World-Class Management Techniques and Philosophy

It can be argued that organizations, employees and the whole gamut of industrial management are in a state of flux as a result of dynamic global phenomena. Over many years, a wide range of management tools and techniques have been developed and are available to give organizations a wide choice of tools that will enable them to achieve their goals. Management can now be considered a science rather than an art. Managing an organization by fortuity and serendipity and maintaining the status quo are approaches of the past. The fundamental concern of successful management is to be able to create an enabling environment that will motivate employees who eventually will become more committed to their job and the organization in general, and hence, increase their productivity.

Several attempts have been made to involve employees in the decision-making process of some companies and to garner their support on major issues as well as to get their opinion on such issues as work improvement, productivity, and quality as well as to give ideas leading to new initiatives. History recalls several management techniques that have had varying benefits, many of which have lost their significance because they lack relevance to prevailing circumstances.

In the 1950s, Peter Drucker introduced Management by Objectives (MBO), which started the revolution in management thinking. This technique was popularised by people such as Odiorne and Humble in the 1960s. Drucker believes that goal setting should become a mutual exercise between management and subordinates. The needs of subordinates should be considered so that they will be able to influence their own work objectives, giving them control over their work environment. MBO was, however, concerned with the day-to-day operational targets and did not achieve its real goal.

Quality circles were introduced in Japan to encourage employees to analyse and solve problems. Job enrichment enjoyed some popularity in the 1960s and 70s. Hertzberg, the proponent of this theory, believed that the job would be enriched if it were designed to provide greater scope for achievement, recognition, challenge and advancement for growth. However, a major weakness of job enrichment is that it claims to be the only way to increase motivation. On the contrary, theorists like Christopher Molander (1989) believe that no theory of human needs tells us what motivates an individual or group of individuals. Transactional Analysis (TA) was another technique used by many companies in an attempt to improve the quality of management and supervision through a greater understanding of the nature of relationships. It was, however, providing insufficient 'real world' problem-solving activity to be successful as a total strategy on its own.

Some philosophies are embedded in techniques. For example, quality circles can ideally operationalise Mc Gregory's Theory Y. Over the years, some of the techniques have become less effective due to poor implementation. Management by Objective (MBO) became entangled with bureaucracy and all its paperwork, which was the cause of much frustration for the average manager. It was also criticised as being too concerned with day-to-day operational targets. Several of the other techniques and philosophies underwent extensive scrutiny and critique and practical experience and application have disproved some of them.

Management techniques and philosophies often provide useful tools for the general management of enterprises, particularly regarding the development, deployment and maximum utilisation of human and material resources. Nonetheless, other complementary quantitative management tools are required, specifically to aid good decision-making. The foundation of management science sometimes referred to as 'operation research', was laid during the scientific management revolution of the 1900s.

Frederick Taylor pioneered this revolution, which gained momentum after World War 11 when specialists (mathematicians, engineers and management scientists) attempted to solve problems using quantitative approaches to decision making. Among the approaches are linear programming, network models, project management (PERT / CPM), inventory models, decision analysis, forecasting, computer simulation and decision and utility analysis. These are important decision-making tools that can be applied to specific circumstances in light of the complex nature of business decisions that have to be made daily. Moreover, the cost of bad decisions can be exorbitant and should be avoided. These tools offer hope for management which is required to make decisions of a quantitative nature.

The five world-class management techniques mentioned earlier can be applied in the context of SMEs. It is not the purpose of this exercise to attempt to add to the existing detailed volume of information pertaining to the various techniques but to take an overview of the different topics to examine their merits and applicability to small developing states.

Production Planning and Control Techniques

In a world where there is uncertainty regarding the future, almost every business decision is based on a forecast of future conditions and events. Forecasting reduces the area of uncertainty that surrounds management decision-making concerning investment, production costs, profit, sales, and so forth. Production planning and control allows the organization to arrange a set of organised activities whereby resources (input) are transformed into finished products in the form of goods and services. Some of the major functions of production planning include forecasting, product design, process planning, routing, material handling and scheduling.

Through proper planning and control, the organization can benefit from improved efficiency, reduced production costs, uninterrupted production flow and timely delivery. To effectively execute a production plan with control mechanisms, several tools and approaches are required. The objective of such exercise is to improve efficiency, reduce cost and waste.

Concurrent Engineering

Concurrent or 'simultaneous' engineering was initiated to create a linkage between design and manufacture; thereby incorporating life cycle values (apart from the product functions) such as aesthetics, serviceability and recyclability into the early stages of a product. In other words, the manufacture of a product is undertaken simultaneously or in parallel with the development of the design and engineering team. CE is concerned with minimizing the time needed to bring a new product to the market. This process involves teamwork, at the center of which is the desire to satisfy customer needs. Products are developed on sound engineering work, design parameters, quality control and assurance and marketing.

Concurrent Engineering is an integrated execution of process management, design, manufacturability, and automated infrastructure support. It is capable of responding to high quality, speed and cost requirements. Several departments within an enterprise are involved in developing the design of the product so that the customers' requirements are adequately reflected in the finished product. Problems related to the manufacturability of the product are addressed at the design stage. This approach increases the opportunity to get it right the first time and reduce cost and waste. Concurrent Engineering brings other benefits to world-class organizations including (i) it reduces time in bringing a product to the market, (ii) it reduces design changes and iterations, (iii) products are more manufacturable and serviceable, (iv) The rate of production is increased because the process is well defined, documented and controlled. As far as developing countries are concerned, Concurrent Engineering can reduce the negative effects of trial and error which so often beset the growth of small enterprises.

The problems associated with some products manufactured in small developing countries have their genesis at the design stage. This integrated approach makes the customer the focal point. It is also cost-effective in the long term and enables a company to respond expeditiously to market changes.

Continuous Improvement (Kaizen)

Kaizen is a Japanese word which means improvement. Continuous improvement is an ongoing exercise involving everyone including managers and workers. Continuous improvement should involve people at all levels of the organizational hierarchy and that improvement should be made daily in selected areas of the company. The kaizen philosophy encapsulates most of the Japanese practices that have achieved worldwide fame. In effect, improvements are contingent upon the implementation of such tools as zero defects, total quality control and quality circles.

Kaizen can also be described as a teamwork approach to improvements. The establishment of kaizen teams is encouraged at various parts of the company to attack waste of any kind in the work area. It is a world-class technique because it involves everyone in the company. Against this background, everyone must be encouraged to forward ideas for improving performance. However, if management is not receptive to ideas, they can do much harm in promoting world-class philosophy. Within a Kaizen group, all sorts of initiatives can be taken, and no topic is someone else's responsibility. Cost reduction and the elimination of waste are topics often discussed. Improvement can be categorized as (i) small improvements made in the status quo due to ongoing efforts, (ii) innovation, which involves drastic improvement in the status quo resulting from a large investment in new technology and/or improvement in existing ones."

Small developing states can benefit significantly if continuous improvement (Kaizen) is encouraged. It will ensure that waste is eliminated and cost is reduced in critical areas that will ultimately impact performance. To achieve world-class status, small enterprises need to encourage teamwork in every area of their manufacturing pursuits. Moreover, the question of quality will be addressed at a later stage.

Lean Production

The current complex and competitive business environment requires companies to make continuous improvements to all aspects of their production processes to keep up with the competitive demands. Lean manufacturing was introduced to improve the whole gamut of the production system by improving productivity, reducing waste, managing space, improving delivery and so forth.

The term "lean production" was introduced by the Massachusetts Institute of Technology to revolutionise and modernise mass production manufacturing systems. The application of lean production entails the following:

(i) Training of workers to be multiskilled;
(ii) Developing cross functional teams;
(iii) Building integrated communications networks and supplier partnerships;
(iv) Introducing highly flexible automated machines to provide a wide variety of products at the least cost.

Proponents of the lean production system think it would be counterproductive to make workers work harder without giving them the tools to work smarter. The system is therefore built on workforce empowerment and participation. In comparing lean production to mass production, "production is rationalised by eliminating unnecessary buffer stocks and idle time by revising cycle times in direct relationship to production needs and running machines at full capacity and giving workers a range of tasks to perform."

Lean production involves a systematic and continuous search for non-value-added activities and sources of waste with a focus on quality and cost. Within the production system, there are new tools and techniques to cut costs, improve quality and promote lean practices. The strategy in a lean production system is designed to eliminate or reduce waste of all types including space, materials, time, energy, inventories and defects. The following is a description of the various wastes that are common in most business enterprises which can affect productivity and competitiveness.

1. Transportation: unnecessary movement of products and materials during the manufacturing process;
2. Inventory: too many products and materials that are not being processed and remain as excess in storage;
3. Motion: unnecessary movement (like using the cell phone for unrelated work, walking, bending) by workers;
4. Waiting: time wasted by workers waiting for the next steps in the process.
5. Overproduction: producing more products that are demanded by the market.
6. Defects: reworking or scrapping of products that are not fit for the customer's use.
7. Non-utilized talent: underutilizing the talents, skills and knowledge of a team of workers.

To achieve this goal, employees are being trained to identify waste of different types and develop ways to eliminate them. For example, workers are trained to maintain and repair machines, thereby minimizing downtime. Managers serve as mentors and coaches in these efforts, which are considered part of every employee's job responsibility. Employees primarily learn the continuous improvement process based on "learning-by-doing." Like many world-class techniques, lean production involves the whole workforce in eliminating waste. This feature aptly characterizes lean production as a world-class manufacturing technique. It emphasises equipment flexibility and people's commitment to total quality as well as devising ways of responding to consumer needs and competitors' activity. In a lean production process, inefficiencies are identified and addressed promptly.

Lean Production offers an ideal tool that can be used by small manufacturing enterprises. Like many world-class techniques, the emphasis is on waste elimination through full worker participation. The principle of lean production is also about delivering the highest-quality product at the lowest cost in a timely manner. Although the practice of lean production has its origin in the automobile industry, the practice has widespread applications. For enterprises in small states to be competitive, the reduction of waste of all kinds must be a priority for competitiveness and efficiency. The good news is that in small manufacturing plants producing simple products, it may be easy to identify all the areas that must be changed to create a lean system. No matter what business is being pursued, there are benefits from the guiding principle and techniques specific to lean manufacturing. In other words, lean manufacturing can be applied to cut waste in the areas of production, customer services, assembly and marketing.

Total Productive Maintenance (TPM)

Like any other gadget in life, machines and equipment have a life span. However, that life span can be shortened or optimized based on proper care and attention that is provided in keeping with prescribed maintenance guidelines. The effectiveness of a facility depends on the smooth functioning of its machinery and equipment. The central goal of an enterprise is to minimise or eliminate downtime caused primarily by equipment failure. The Japanese learned about preventive maintenance from the U.SA and they have since developed and improved the preventive maintenance programme into what they presently call Total Productive Maintenance (TPM). The phrase Total Productive Maintenance was first coined in 1961, inside the Japanese company called Nippondenso. At that time, a supplier to the automotive industry carried out a project named 'productive maintenance with total employee participation.' The development stages of TPM were recalled as follows:

1. In the 1960s, Preventive Maintenance - establishment of key maintenance functions
2. In the 1960s, Productive Maintenance - reliability, maintenance and the economic efficiency of plant design were recognised as important.
3. In the 1970s, Total Productive Maintenance - the development of a comprehensive system involving total employee participation to achieve productive maintenance efficiency.

TPM involves everyone from the operator to the maintenance engineer. It was a joint effort to improve the overall effectiveness of the equipment. This philosophy has become the central theme of Total Productive Maintenance (TPM). The process is made up of small multidisciplinary teams, that 'own' machines or production lines. Attention is given to autonomous maintenance, preventive maintenance, training, security and standardization of working methods with the ultimate goal of zero errors, zero work-related accidents and zero losses.

Preventive maintenance is performed periodically to reduce the incidence of equipment failure and the costs associated with it. These costs include disrupted production schedules, idled workers, loss of output and damage to products or other equipment. Preventive maintenance can be scheduled to avoid interfering with production. In the past, maintenance of equipment and machines was done in an arbitrary and unplanned manner, especially when a problem was uncovered. Ideally, preventive maintenance will be scheduled to take place prior to the failure of machines, so that the operational time of equipment could be maximised between scheduled maintenance activities.

Production managers were required to find a balance between preventive maintenance and breakdown maintenance that will minimise the company's overall maintenance costs. If maintenance work were done only in a reactive manner, after a breakdown occurs, there would be an increase in the overall cost of repair. There are several hidden costs such as lost production, the cost of wages while equipment is not in service, injuries, and damage to other equipment and facilities or other units in production. The costs associated with these side effects can be minimized by TPM. There is, however, a point at which the cost of preventive maintenance exceeds the benefit.

The decision of how much maintenance to perform involves the age and condition of the equipment, the complexity of the technology used, the type of production process and other factors. For example, managers would tend to perform more preventive maintenance on older machines because new ones have only a slight risk of breakdown and need less work to stay in good condition. It is also important to perform routine maintenance before the beginning of a large or important production run.

In TPM, production employees are trained in both operating procedures and routine maintenance of equipment. They perform regular inspections of the machinery they operate and replace parts that have become worn through extensive use before they fail. Since production employees spend so much time working with the equipment, they are likely to pick up small signals that a machine needs maintenance. Among the main benefits of TPM is that employees gain a more complete understanding of the functioning of the system. TPM also gives them increased input into their own productivity and the quality of their work. Equipment failure can be considered the main focus of a TPM programme. It can also be attributed to losses in time and production that have exacerbated the cost and limited the effectiveness of equipment in plants. The following are contributory factors that have affected production optimisation:

(i)   equipment failure and breakdown;
(ii) setup and adjustment downtime;
(iii) idling and minor stoppages;
(iv) reduced speed;
(v)   process defects;
(vi) reduced yields.

To devise a strategic and effective maintenance programme, consideration should be given to the role of each of the following categories within the overall planned maintenance programme:

(i)   Preventive Maintenance is an attempt at a maintenance-free system where equipment are easy to repair and safe to handle. This will ensure minimum maintenance cost to an organization. This system involves daily or weekly inspection of equipment by users. Activities in this system of maintenance involve cleaning, adjustment, oiling and minor replacement. It requires the maintenance people to adjust equipment according to stipulated maintenance standards. It also requires regular inspection for any abnormality, replacement of worn parts and repair of damaged parts.
(ii) Breakdown Maintenance is initiated to effect repairs on equipment after failure. This will require that spare parts are immediately available to avoid prolonged downtime.
(iii) Corrective Maintenance is an improvement in equipment design to prevent failure. At the operational stage, maintenance can be planned or unplanned.

The fundamental concern of a TPM programme is to increase its effectiveness to its highest potential and maintain it at that level. This can be achieved by understanding the above factors and devising a preventive maintenance programme to address them. TPM should also be concerned with the prevention of failure of all equipment before it occurs. Signs of failure must be detected early, and actions must be taken promptly.

Ultimately, total productive maintenance aims to increase productivity, improve efficiency, reduce downtime and set-up time and loss of speed. It is considered a waste if machines are not performing to their optimum. To be effective, operators need to be trained as well as have the cooperation of maintenance specialists. Designated skilled technicians should be included on the production staff to monitor the performance of equipment.

One of the common characteristics of machinery and equipment in enterprises in small developing states is a high failure rate resulting in significant downtime and related costs. In many cases machinery are antiquated and spare parts are not available. Success in implementing a TPM system in small states will require a change of culture to include the entire workforce and increased demand for more training opportunities leading to continuous improvement. There is often a shortage of technicians with the requisite in-house skills to effect repairs to machinery. As a result, technicians are often brought in from another country at exorbitant costs. This is often the result of poor negotiation at the technology transfer level when technology (embodied in machinery) is transferred to a smaller less developed country. The providers of such technology should be contracted to provide after-sales maintenance services as part of the technology transfer package. Too often small states are left with non-functioning equipment for prolonged periods, which is another form of waste that has a cost associated with it. More often than not, there is no "Total Productive Maintenance Programme" to address the matter of machine failure, downtime, waste and optimal machine and equipment performance.

Chapter 8

Managing Quality to Improve Competitiveness

The concept of quality has several connotations. Essentially, it refers to the set of inherent properties of a product or service that provide satisfaction to the stated or implied needs of customers. These stated qualities of a product allow a good or service to be compared with any other good or service of its kind. Every customer has a mindset or perception about a product. That perception promotes the notion that a specific good or service can provide the fulfillment of one's expectation.

When the quality delivered by a product or service surpasses the expectation of the customer, there is often an astonishment or 'wow" effect. A judgement of this nature could be very subjective because it is sometimes difficult to measure exactly what will meet the basic needs of customers consistently. In today's competitive world, quality is often determined by several factors that hitherto were not considered important. Product safety, serviceability and ease of use are factors that determine product quality. These factors must of necessity be incorporated into a product at the design stage and managed throughout the production of the product. Quality control becomes important because it ensures that quality is delivered and that the customer has no regret about purchasing the product.

It is the desire of every customer to buy the best quality goods that will perform the desired function for which it is intended at an acceptable price. It is for this reason that management practices are so dynamic, giving rise to the adoption of new systems that are designed to produce the best products that will meet the changing needs of customers. In some instances, there are variations in the concept of "quality" which consequentially influence the manufacturing systems and practices that dictate how products are designed and produced in keeping with product quality.

A few decades ago, quality was defined as "conforming to the standards and specifications of a product." The main quality practices adopted by industries were to have a standardized system of quality, inspection and rework. Deming emphasised that the central tenet of quality is the fulfillment of the requirements of customers and thereby satisfying them. Hence, quality was widely perceived as one based on "customer focus." Every enterprise wants to pursue the goal of satisfying customers' needs and expectations.

Consequently, they have developed several methods to identify those needs and expectations so that they can design products with inherent characteristics to address those needs. In this regard, several approaches including market research, customer interviews and surveys are conducted on a regular basis.

Over several decades, the concept of quality has changed drastically. Today, quality has been intricately linked to customer focus. Hence, the buzz word associated with quality is "customer focus" and "customer satisfaction." As a result of this new focus, companies spend an enormous amount of resources and time trying to identify those factors that will reveal customers' needs and expectations. There is, however, a new thinking that industry should not only consider customers' physiological needs but also focus on those products that appeal to customers' pleasure, excitement and other experiences. The story of how Apple has created innovative products that appeal to the delight of customers is well documented and intriguing.

A term such as "customer delight" is becoming one of the underlying elements of the quality concept. The concept of delight is an expression of emotion exuding joy, euphoria, and surprise, which can only be achieved when customers' latent needs are satisfied. Changes in the concept and notion of quality will lead enterprises to devise an innovative strategy for product development. They will thereby re-engineer their existing quality system to develop the innovative quality attributes of products and services, and in so doing, retain and attract customers.

Quality is one of the main aspects that is used to assess the cost and efficiency of a product, but in some fields such as engineering, quality deals with safety, structural integrity and serviceability. It is for this reason that the science of quality control was introduced and should be understood clearly to manage quality effectively. Quality control, or QC for short, is a process by which entities review the quality of all factors involved in production. It is also done during the production or construction process.

Quality Assurance and Quality Control Approach

The development of the "total quality management" (TQM) philosophy in the 1980s was an important landmark, which caused the change of new concepts and the re-engineering of the quality management system. A quality system encompasses quality assurance and control and includes other factors that enhance product quality. The term "quality assurance" is often used interchangeably with quality management systems or quality control (QC), but they are not the same. Quality assurance typically covers the entire gamut of the quality system, whereas quality control is a subset of that system. On the other hand, a quality system involves a wider range of issues and is more complex than quality assurance and quality control.

Quality assurance (QA) can be defined as the process of ensuring that a product or service meets specified standards. It is rooted in the manufacturing industry and includes vendors, materials and material handling and the production processes. The establishment of a quality system specifies a set of systematic activities that should be undertaken or implemented consistent with the quality requirements of a product or service.

On the other hand, quality assurance is primarily used to ensure that each product is produced to the highest standard. It generally involves a rigid and systematic set of activities including measurement and monitoring of the various processes and comparison with a specific standard for compliance. There is also a feedback mechanism that is designed to prevent errors. Quality assurance is critical in ensuring that a product meets the established requirements.

In so doing, all the checks and balances are intertwined to ensure that the product fulfills those criteria that guarantee customers' satisfaction. As a corollary, any compromise in the quality assurance system would undermine customer satisfaction and the confidence that a product will deliver the values for which they pay.

Quality control is that aspect of quality assurance that focuses specifically on meeting a set of quality requirements including a range of activities from technical operations to product inspection. Quality Assurance involves a combination of planned activities and systematic actions that are critical to enabling an entity to fulfill its quality obligations. A rigid and credible quality assurance system allows customers to build confidence in the product they buy, even if they cannot witness the various operations that contribute to the development of the product. This means they can trust the quality assurance process to deliver on those attributes imbued in the product that provide customer satisfaction. Quality assurance should be ascertained by testing the products or services against prescribed standards to establish the capacity to meet those standards. It is important to note that quality assurance does not control quality but establishes the framework by which quality will be controlled. The most effective means of quality assurance is to build elements of assurance into the process. For example, establishing records, documenting specifications and reporting reviews are critical to ensuring that quality protocols are adhered to.

An enterprise can also encourage customers to visit manufacturing plants to observe procedures as described in the operations. This can assist in building confidence among customers. Audits of the quality control process should be carried out regularly and recorded in the quality manual to measure the system's effectiveness and thereby ensure maximum compliance with the established procedures. With the advent of free-market access, small developing states are beginning to take the issue of quality control seriously. Quality assurance was, and in many cases is still taken for granted by many small firms. With more consumer awareness programmes, consumers will get a better understanding of the production systems vis á vis its nexus to established quality standards.

Every manufacturing enterprise should establish a realistic and credible quality assurance system that will address the salient issues that affect quality. To do so, a quality manual should be developed which will comprise several management documents. Among other things, Standard Operating Procedures (SOPs) detailing all specific operations and methods, including sampling, logistics, analysis, calibration of equipment, production of reports and interpretation of data are tools to enhance quality assurance protocol. An internal reference manual should also form part of the literary repertoire for the particular procedure and should detail every relevant step.

Method SOPs may be accessed from international standard bodies including the International Organization for Standardization (ISO), British Standards Institute (BSI), American Standard Technical Method (ASTM), or from the instructions that come with the test kit where a commercially produced method is used. Such SOPs have the advantage of not requiring verification and thereby save time in writing "in-house" SOPs.

Equipment SOPs should include methods and frequency of maintenance, cleaning, calibration and servicing. Method SOPs should include all the information necessary to carry out the procedure without reference to other documents except for fully documented SOPs.

Testing is an important aspect of the quality assurance process and should involve every team member. Each member of the quality assurance team must understand the requirements of his/her task which must be performed in accordance with the accepted guidelines and protocols. During this process of testing, physical product teams will continuously test a product until it physically breaks or fails.

Quality Circles

A quality circle is a tried and tested approach to quality management. It is used to define, identify, analyse and solve quality or performance-related problems. This practice originated in Japan in the 1960s, in response to industrial problems such as strikes, absenteeism, excessive sick leave, and other grievances from workers. This practice, also known as quality control circles has become an integral part of enterprise management. QC promotes participatory management techniques regarding a companywide quality system. The approach involves a small group of workers (usually between four and ten) from within a department who meet voluntarily weekly for about an hour (on company time) under the leadership of the leader to address quality problems that occur within their department." Implementing a successful quality circle programme will require a new culture to be fostered within the organization.

For QC to succeed, top management must provide support to their subordinates and demonstrate commitment. Persons in the group should be more or less homogenous with similar educational backgrounds and experiences. Small groups are more effective than large groups since large groups can inhibit participation by everyone. Quality circles have had successful results in many countries around the world where it is being practised. It can be applied in several industries and departments.

Among the main features of quality control are that any programme should begin in a very small way and should be voluntary and unbureaucratic. There are benefits to be gained both for the individual and the organization. The members gain confidence and competence in problem-solving and there is generally a change of attitude which can lead to greater commitment to the organization.

QC is a management technique worth implementing in manufacturing enterprises of small developing states. Enterprises in these states are sometimes rife with indifferent work attitudes and problems between workers and management. The culture of most companies will have to be changed to encourage such a technique. Management should encourage the involvement of subordinates in all quality matters and demonstrate commitment by allowing the technique to work successfully.

Statistical Process Control (SPC)

Statistical Process Control (SPC) is a tool used for monitoring and controlling manufacturing processes. Irrespective of size, it is useful in monitoring batch or continuous production by sampling at set frequencies. An SPC system provides a warning when there is a potential quality problem so that it can be averted through the necessary corrective action. Of course, poor quality products are deleterious to any organization. SPC mirrors statistical quality control which was developed to check samples of products that are produced in batches. Over many years, the process has been simplified by removing the mathematical component and focusing on a key feature such as control charts that make it simple to understand and use.

Statistical Process Control is also used for reducing variations in products, deliveries, processes and materials. It is important to note that variability is one of the main reasons for quality problems. The fundamental reason for SPC is hinged on the premise that all things vary, hence, statistical methods of quality control must be able to measure and gain an understanding of the causes of the variation so that it can be managed.

SPC relies heavily on information which forms the basis for decision making. The process is therefore heavily dependent on the information system for support. Among the basic SPC tools used for analysing and interpreting data are flowcharting, check sheets, and histograms. SPC is a necessity in every organization, not least in those of small developing states where resources are scarce and the cost of production is generally high due to variations in materials and processes. If variation can be managed successfully, it will enable enterprises to reduce waste and increase efficiency and productivity.

Total Quality Management (TQM)

In the contemporary dynamic environment, many traditional business organizations have not demonstrated that they have the capacity to cope with the challenges of modern corporate operations. Advanced managerial concepts emerged in many world-class companies as mechanisms for reaching their objectives. One of these mechanisms is Total Quality Management (TQM). TQM is an approach that promotes market sustainability of organizations by improving their efficiency and flexibility and putting them on a trajectory to competitiveness. The implementation of TQM enables organizations the ability to rapidly adapt to the changes in the environment, which enables them to become competitive in the global market. Quality management systems like the ISO 9001 standard in many organizations represent additional stimulation for further TQM development. ISO 9001:2008 is a quality standard that has become a commonly accepted management tool and is universally applicable to many situations, particularly where advanced management systems are in effect.

The adoption of TQM is voluntary and is oriented to various aspects of management and does not delve into the merits of specific operational activities. The system is based on a declaration of a quality policy, on the definition of specific objectives and further continuous improvement through measurement and monitoring of customer satisfaction, process performance and quality objectives. TQM philosophy represents an integral part of the creation of a climate in which employees are encouraged to locate quality problems and find solutions for them.

Total Quality Management is an approach aimed at improving the effectiveness and flexibility of businesses. For TQM to be effective, the entire organization including every department, every activity and every person at every level is involved in the total quality of the organization's pursuits. The total involvement of every worker in attacking waste makes the technique world-class. Workers must be held accountable for failures. From the chief executive to the shop floor worker, everyone is encouraged to accept personal responsibility for getting it right, the first time and every time. This concept emphasises the prevention of errors, unlike the traditional thinking that some errors will occur, then they can be identified and dealt with.

If everyone were to be involved in total quality, management must take the lead by demonstrating its seriousness about quality. Management must also demonstrate its commitment to the rest of the company by promoting commitment and constant improvement, a "right first time," philosophy that involves training people to understand customer-supplier relationships. Quality is recognized as the most important competitive tool in meeting customers' needs and it has been used strategically to win customers and enhance competitiveness. When maximum attention is paid to quality, it improves performance in reliability, delivery and price.

The TQM philosophy can be implemented successfully in manufacturing organizations of small developing states as a competitive tool. However, people must be willing to change the old approach to doing things. Top managers will have to create a new culture within the organization and adopt as a strategy an approach that characterizes TQM as a world-class technique including total involvement of every worker, encouraging and promoting actions that improve efficiency and eliminating waste by avoiding errors.

Emphasis will have to be placed on three basic elements (i) a good management system, (ii) using tools such as statistical process control (SPC), and (iii) teamwork. Training should therefore be given high priority so that workers are equipped with the necessary skills and knowledge to implement the programmes.

Total Quality Control

Total quality control can be used as an effective system for integrating the quality development, quality maintenance and quality improvement efforts of the various groups in an organization. This can ensure that the production of goods and services can be realized in the most economical manner that will satisfy those needs critical to customers' loyalty.

Effective human relations are the basis of quality control. The Japanese adopted the concept of total quality control from Feigenbaum and renamed it "company-wide quality control." Its success in Japan was due largely to three main reasons (i) upper managers personally took charge of leading the quality revolution, (ii) all levels of employees and functions underwent training in managing quality, (iii) quality improvement was undertaken at a continuing rapid pace. Quality control must be all-encompassing beginning with the design of the product and ending when the product reaches satisfied customers.

Another useful approach to quality is Six Sigma. This is an approach geared towards quality improvement which lowers costs and increases productivity and profits through statistical and problem-solving tools that result in remarkable improvement. The tools are applied by trained practitioners known as Black Belts. Sigma is a measurement of quality that determines how effectively defects and variations from processes are eliminated. The system is recognized worldwide as a measure of excellence.

Companies in small developing states that are implementing a total quality control system should provide the necessary framework by having their management leading the way. The appropriate training should be provided for its employees and consistent quality improvement must be pursued using a combination of methods. The focus should not only be on satisfying the customers' needs but to exceed them. This goal can be achieved by eliminating waste, reducing cost and improving efficiency and quality. The entire organization including employees and departments should be involved in the exercise.

Total Employee Involvement

It can be argued that people are the most important asset of an organization. For an organization to become world-class, it must maximize the involvement and contribution of its employees. Machines and technology may empower an organization to forge ahead in the market, but the catalyst of organizational success is its employees. Management should devise a framework for a participatory approach by involving subordinates in the decision-making process. This will promote organizational effectiveness and enhance employee satisfaction.

Those organizations that take an approach of employees' inclusiveness in decision making and the like, are more likely to solve most of their problems as it is said, "excellent companies fire on all six cylinders."

Continuous improvement within an organization can be enabled by employees who are motivated to participate in the decision-making affairs of the organization. The importance of motivation cannot be overemphasized. Motivation can increase the involvement and dedication of the employees in the process of transition from ISO 9001 to TQM.

Employees should be continuously encouraged and supported to participate meaningfully in the affairs of the company. Total Quality Management (TQM cannot be effectively implemented without the active involvement of employees. Companies that do not prioritize employee involvement in their strategic planning are often rife with problems of teamwork and communication, failure to meet deadlines and increasing costs in several areas. Employee motivation means incentivizing them to become dedicated participants in those operations of the company that contribute towards the improvement of the TQM practices in a shorter period.

Employees possess more knowledge and creativity than management would often perceive, hence management can evoke those latent talents and other endowments that are essential in building teamwork and accomplishing major quality achievements. If employees were empowered by being given the tools to make decisions that will affect their jobs as well as the freedom and encouragement to contribute to critical organizational issues, the organization stands to benefit from higher productivity and quality products.

Through total involvement of all employees, management can successfully foster a common commitment to quality that is shared by everyone. When quality improvements are reported in the firm, it can motivate other departments in striving to achieve quality targets. For employee involvement to be achieved, management must develop an effective communication process, which will outline the mission of the firm as well as its goals and objectives in a lucid manner. Employee commitment can be reinforced by various remuneration schemes, employee security and performance measures (sales quotas, efficiency measures and output-based incentives).

Total employee involvement is important to every organization irrespective of size and resources, as it strives for efficiency and to become competitive. Enterprises in small states should devise systems whereby employees can play a greater role in the affairs of their jobs. Failure to do so can result in poor quality products, which can undermine the firm's competitiveness. In a regional grouping where there is a common market to facilitate trade among member states of that group, it is evident that there is no motivation to expand businesses beyond regional boundaries. In these circumstances, organizations expend little resources on marketing, technology acquisition, training, product development and other related areas that will allow them to meet the requirements of the global market and augment the opportunity for greater prosperity. Total employee involvement is sometimes misunderstood, or there is naivety in addressing it. This is evident that when there is a failure to train employees to respond readily to new challenges, they are likely to make costly errors.

Benchmarking as a Competitive Tool

Benchmarking involves the continuous process of measuring products, services and practices against the toughest competitors, that are recognised as industry leaders. In essence, benchmarking involves looking at, and learning from others by comparing oneself with them. It must be emphasised that performance and behaviour are not static, they change over time. Benchmarking is a long-term process, which involves the whole organization in search of the best practice outside the company. It does not only involve what is done best, but how it is done. The process helps companies to examine underlying operative practices and success factors and behaviour. General Sun Tzu a Chinese military strategist and philosopher once wrote: "If you know your enemy and know yourself, you need not fear the result of a hundred battles." Benchmarking originated from Xerox Business Systems in the late 1970s. It was used as a tactical planning tool in response to the domination of the copier market by its Japanese counterparts. They experimented with benchmarking in the areas of production logistics (warehousing, packaging and shipping).

A few years ago, a study was conducted in several selected firms on a small island developing state. It was found that the concept of benchmarking was a new phenomenon. As a result, it was not given priority attention as a competitive strategy because the concept was not understood and the competence to conduct the exercise was lacking. Some of the areas that are critical to world-class techniques such as quality, employee involvement, training and elimination of waste have not been given serious attention. As a consequence of this exercise, it was revealed that these enterprises wholly lacked world-class characteristics in their operations and could not compete favourably with world-class competitors.

Genuine benchmarking should be viewed in the context of an organization that is continuously assessing itself by analyzing its performance and internal processes, thereby implementing improvement measures on an ongoing basis. It is important to ascertain the status of the firm's operations and performance on a given set of criteria. This would allow the firm to establish targets for improvement in accordance with a given action plan. Benchmarking can be divided into four groups, viz.

(i) Internal benchmarking – comparison with other parts of the same organization.
(ii) Competitive benchmarking – understanding how competitors operate.
(iii) Functional benchmarking – comparison with other competitors that have the same functional activity, e.g., warehousing.
(iv) Generic benchmarking – a comparison that cuts across functions or different industries.

Benchmarking world-class performance starts with competitive analysis, which focuses on product comparison. However, it goes beyond product comparison and addresses operating and management skills that produce the products. Competitive analysis has its limitations, in that only companies with more or less similar products are examined. On the other hand, benchmarking looks at the best processes or skills wherever they are found.

In an era of liberalised trade, benchmarking has become an important tool for every manufacturing enterprise in small developing states that is striving to attain world-class status. The attainment of such status will ensure that a company is efficient in its manufacturing practices; that management is effective, and there is minimal waste of the company's resources; employees are motivated and involved in decisions about their jobs; productivity is high and customer needs are satisfied. Ultimately, the company can compete with the best in the business in the world.

A firm should benchmark its operations against the best in the business in the world periodically. This exercise should encourage them to take measures to incorporate the appropriate technology as well as other requisite good manufacturing techniques. The use of an effective benchmarking tool can enable firms to benchmark their activities against world-class characteristics. Subsequently, firms should embark on a programme of continuous improvement. This is not necessarily about a firm learning from its mistakes, but an effective benchmarking programme should involve a comprehensive and ongoing programme aimed at identifying gaps between itself and world-class competitors. It is therefore imperative that firms that are desirous of improving their status quo devise a plan of action aimed at minimising those gaps and to keep improving continuously to ensure sustainable competitiveness.

A firm that is being benchmarked should place more emphasis on management techniques and strategies. The most outstanding companies are those with strong management structures and techniques, which can be considered catalysts for the adoption and implementation of other world-class techniques. Failure to implement relevant management techniques that will result in giving an organization a competitive edge may result in a quick exit from the market. Good incisive, proactive and change-conscious management will employ the appropriate technologies, encourage employee involvement and build strategic relationships with suppliers. Some organizations have quasi forms of world-class manufacturing techniques but have not fully implemented them. Other organizations have attempted some aspects of implementation but discontinued them because they lack the resources, expertise and experience. In some cases, there is a lack of knowledge about world-class manufacturing techniques. Some companies had conducted some rudimentary aspects of benchmarking in the past, while others had never conducted the exercise.

Firms in small developing states need to pay more attention to what is happening at the global level, which they should target as their ultimate operating environment. They need to be sensitised to the benefits of such concepts as world-class management techniques, benchmarking and continuous improvement. It must be categorically stated that knowledge of regional competitors, their management techniques, the technology used, and the marketing strategies employed provide no good reason for complacency and lethargy. Most enterprises seem to be satisfied with their present state of progress. Furthermore, it is important to reiterate that size is not a deterrent to achieving world-class manufacturing standards.

Chapter 9

Standards; A Catalyst for Competitiveness

The ISO (International Standardisation Organization) refers to standards as "documented agreements containing specifications or other precise criteria to be used consistently as rules, guidelines, or definitions of characteristics to ensure that materials, products, processes and services are fit for their purpose." A standard is a detailed statement of requirements. It is a set of criteria by which to measure a product or service and is often established by authorized agencies, as well as driven by well-established traditions and customs embedded in documented procedures. It is important to note that quality cannot be achieved without standards.

A standard specification involves an explicit set of requirements that are necessary to produce an item, part, system, or service. Standards are often used to formalize the technical aspects of a procurement agreement or contract. In most situations, a standard test method outlines a definitive procedure necessary to produce a desired result. Global markets require products that are manufactured in one country to be replicated across several countries using standards and specifications.

In many situations, parts can be manufactured interchangeably, particularly in cases where mass production is practised. Products are generally required to conform to specific specifications that must be supported by certification programmes which will ensure that those products meet specific customers' requirements.

Performance-based specifications are often confused with design specifications. Performance-based specifications focus on outcomes or results of the specific goods and services rather than the manufacturing processes. Such specifications provide the required performance criteria that must be delivered by the equipment, for example, the throughput, the pressure, production capacity, etc. Design specifications, on the other hand, outline the process necessary for the manufacture of a product, or how a particular service should be performed to get a desired result.

Standards and Quality Approach to Competitiveness

The processes that influence and drive competitiveness are often multifaceted and dynamic and are consistent and compatible with the prevailing global order. The current market dynamism emerges in response to the global proliferation of new and advanced technologies that have driven the revolution in new product development and corporate strategies.

Standards define the degree to which a set of inherent characteristics reflect customers' requirements and expectations or comply with stated norms, regulations and laws or both. Standards and quality are intrinsically linked, in that, standards are used to codify the technical characteristics expected by customers. The provision of quality goods and services by industry relies on a sound standards infrastructure.

Globalization has engendered deep and pervasive integration of the world economy because of the increase in global flows of information, ideas, production factors, technology and goods. Successive rounds of international trade agreements have systematically reduced trade barriers in developed as well as poor developing countries. Transactional corporations and global buyers impose standards on their suppliers to ensure compatibility between products and processes throughout the global value chain and they use standards to satisfy high customer demands. In light of increased competition triggered by free trade, developing countries aspiring to sustained growth should transit from the dependence on primary products and diversify into value-added manufacturing geared towards the export market. A poor investment climate, small markets and the lack of requisite standards inhibit the development of small firms and their ability to enter global markets.

It is instructive to note that quality is defined by standards. Standards represent a set of requirements and rules developed to ensure safety, quality and good trading practices. Standards can be specific to a country, a region or the international market. Notwithstanding the noble ideals of standards, they are often used as technical barriers to trade, thereby circumventing and undermining the main purpose of free trade among countries. Within small regional trading blocs, standards for specific products fluctuate for convenient reasons and therefore make it difficult for some manufacturers to keep abreast of those stringent, albeit inconsistent requirements. Frequent changes in standards often require a redesign of labels, adjustment to equipment and product content, production processes and the like.

In recent years, the regulations governing some standards do not only relate solely to the product content but also to the environment within which products are developed, more so in the food industry. Small developing states often regard standards requirements, particularly international standards, as burdensome because of the exorbitant costs of meeting those requirements, both in terms of equipment and personnel for surveillance.

The public sector is generally limited in its provision of oversight services for standards and quality. Inspection departments generally have limited human and physical resources and support infrastructure and facilities. In many cases, laboratories for testing and monitoring are limited or non-existent. Because of limited capacity, services for testing and standard development are in great demand, hence, the costs of these services are prohibitive for smaller producers. These support systems are essential to enable the industrial sector to address the technical barriers to trade and to meet the certification requirements necessary for penetrating export markets. These weaknesses also increase the vulnerability of regional economies to unfair trading competition because some countries lack the capacity to regulate their imports, thereby placing an undue burden on local producers.

It is well understood that the global marketplace is dynamic and poses new and varying challenges for small island states. For enterprises in small developing states to enter and compete in global markets, they must first establish the requisite technical and administrative mechanisms and infrastructure to augment their competitive position. These mechanisms should be informed by an understanding of the systems and procedures that are required for conformity to regulations and guidelines germane to international quality systems. Among the preconditions for quality protocols and guidelines are Hazard Analysis Critical Control Point (HACCP), Codex Alimentarius, the ISO series and Six Sigma Quality.

Adherence and compliance to established quality standards and norms can be regarded as unfamiliar to many SIDS because they have never been enforceable preconditions to trade. Moreover, the exorbitant cost of establishing the necessary systems and mechanisms to facilitate compliance is prohibitive. The banana industry is a case in point. For many decades, many small island developing states like those in Africa, the Caribbean and the Pacific (ACP) grew and exported large quantities of bananas to the United Kingdom, without having to comply with stringent standards requirements.

During the decades of the eighties and nineties, farmers were preoccupied with and encouraged to 'grow more bananas' and not necessarily "grow better quality bananas." The subsequent removal of the protective trade regime for bananas from the ACP states, allowed more bananas from other countries notably Latin American countries, to enter the market that hitherto gave protection to bananas from former colonial states in Africa and the Caribbean, thereby providing greater competition. A new competitive environment emerged, which took the farmers by surprise and in the process created numerous challenges for both the farmers and the governments by extension. In subsequent years, the emphasis has been on quality, and farmers had to be educated and reorient their strategy accordingly.

There are many lessons to be learned from that experience. Policymakers and entrepreneurs should not allow a worst-case scenario to occur before the necessary systems are instituted. Standards and quality control measures should be perceived from two points of view (i) the protection of the consumer from substandard products and (ii) to encourage entrepreneurs to develop better products and thereby become competitive and gain market share. The decision of whether to comply goes beyond the cost factor. It has become an imperative as well as a prerequisite towards becoming world-class and competing in the global market.

The competitive nature of the business environment in a dynamic global economy places a demand on enterprises to prioritize efficiency and competitiveness. This can be achieved through detailed planning and management of production activities and the allocation of scarce resources. In most cases, this approach to industrial development and manufacturing necessitates new and advanced technologies and management systems that will enhance efficiency and competitiveness.

In the 1980s and 1990s, many third world countries were bound by debt commitments to multinational agencies towards which they channel a significant proportion of their GDP. This allowed governments little fiscal space and wealth creation became paramount to keeping vital state machinery functioning. Governments were thereby required to examine various options for income generation. In the face of limited resources, the efforts of many governments are sometimes limited and ephemeral. In such circumstances, private sector involvement is necessary and desirable.

Production efficiency must respond readily to market demand as well as technological innovation, and new corporate strategies aimed at gaining market share. A critical problem facing industrial enterprises in small developing states is their inability to improve their productive capacity and efficiency and ultimately their competitiveness in the face of market dynamism. Strategies need to be devised to assess and improve the productive capacity of enterprises so that they can increase market share and revenue. As was stated earlier, the use of relevant manufacturing techniques comparable to World-Class Manufacturing standards is one approach that may address many areas of concern. This approach should result in managerial and technological improvements that would augment the capability of enterprises and therefore exploit trade and market opportunities.

The main focus of the so-called implicit industrial policy for many small island developing states has been import-substitution, which has resulted in inefficiency and low production in the industrial sector.

It was envisaged that through import substitution, local industries would enjoy a minimum level of protection through high import tariffs, quota restrictions on imports and extensive fiscal incentive schemes that discourage competition. The avoidance of competition has forced industries to be predominantly in-ward looking. In such economies, the export sector is generally penalized relative to the sector producing for the home market. Such circumstances usually result in low sales volume on the global market due to a limited number of products that can compete in such markets.

Changes in the global economy such as deregulation and privatization, the formation of new trading blocs, the global integration of financial and capital markets, mobility of capital and technology and increasing acceleration in the development and deployment of knowledge have resulted in rapid product development and or obsolescence of same. Increased training in advanced technology has served to spread economic power among a diverse number of countries, thereby increasing inter-regional competition. These developments represent enormous challenges for SIDS as they seek to widen their industrial base to involve innovative products against the backdrop of technology-based products that account for burgeoning market share, as opposed to conventional price competition.

With the abandonment of the NIEO mentioned earlier, it was envisaged that restructuring the economies of developing countries would put them on the path to reducing poverty through monetary and fiscal measures, together with the liberalisation of the market. However, this approach has failed to produce growth and equity. Small developing countries need a strategy that will ensure that the government maximizes the revenue generated from their production of primary commodities. There is always a widespread concern that failure to attract adequate funding for development projects, which could provide jobs for the poor, could lead to poverty, drug smuggling, and money laundering.

In the face of intense competition, appropriate measures are therefore necessary to develop the capacity and capability to keep abreast of the changing dynamics of competitiveness and the new realities of the global market. It is therefore in the interest of small developing states that they develop industrial policies that reflect their own economic realities and that of the economic bloc they represent. There are lessons to be learned from those countries with similar historical and economic backgrounds and experiences that have experienced substantial economic growth. Enterprises should evaluate their own position regarding the global scheme of things and take cognisance of those gaps that exist between themselves and their competitors and then devise strategies to address them. They should also be amenable to new technologies and techniques that can improve their efficiency and productivity.

How Standards Impact Quality

Over the last several decades, changes in global trade flows have enhanced the role of quality and standards in economic development. A shift in trade composition has given greater importance to manufacturing. Although manufacturing has fallen in some developing countries, it is still one of the most important export sectors in many small developing states.

Small and micro-enterprises, as well as small farmers can improve their social status and subsistence level of production by adopting relevant standards or by making minor adjustments to their product offerings, which have led to new market opportunities. SMEs in developing countries lack the knowledge, expertise and resources to adopt and implement high-quality standards. Unfortunately, enterprises that are unable to meet these standards find themselves excluded from the global market. Competition primarily based on quality often leads to a more sustainable competitive advantage than that based exclusively on price.

Standards have become imperative in every manufacturing enterprise. No business enterprise irrespective of its size and resources can survive in the market if its product, processes and general code of manufacturing conduct were compromised. Quality systems ensure that customers receive products and services that meet and surpass their expectations. Enterprises in small developing countries need to embrace the concept of Total Quality Management (TQM) which emphasises the philosophy and style of management that gives everyone in an organization the responsibility for delivering quality to the customer. This philosophy would work harmoniously with the Value Chain method, particularly in value-added activities.

Adherence to a quality system is one of the pillars of global competitiveness. The cost of establishing accredited standard infrastructure, assessment of laboratory services and the implementation of the required system of standardisation has become an exorbitant exercise for many small enterprises. It is for this reason that adherence to standards and quality could be a challenging experience for micro and small enterprises. A common concern expressed by the management of enterprises is the cost of implementing a quality system such as HACCP. In the absence of inadequate infrastructure, some of the manufacturing and processing activities are conducted in partially dilapidated buildings and under poor environmental conditions. Packaging and labelling are unattractive for the most part and equipment are archaic and do not give accurate results because they are not properly calibrated.

Product quality can also be compromised because of poor quality raw materials from suppliers because there is no formalised agreement and relationship between suppliers and manufacturers on matters of quality and consistency of supplies. Furthermore, this problem can be further exacerbated if there was no monitoring system to monitor the quality of raw materials being supplied to ensure that they are not of substandard quality. The bureaus of standards (or appropriate agencies) should assist with some testing, notwithstanding they function primarily as regulatory bodies. In the case of the smaller territories, the government may be able to assist by establishing national laboratories that will provide support services to these small enterprises at subsidized costs.

Despite the issue of cost, small enterprises should endeavour to pursue a policy of achieving excellence by adopting international quality standards including ISO 9001, HACCP, Good Manufacturing Standards and EUREP-GAP, as well as meeting other stipulated standard requirements necessary for trade with selected countries and other trading blocs. Some of these protocols can be implemented on a phased basis which will allow the enterprise time to leverage the resources to implement other aspects of the quality system.

Quality should be promoted as the identity and hallmark of competitiveness in products and services with strict adherence to environmental standards, labour laws and other normative codes of industrial conduct. Small states should aggressively pursue the development and harmonization of standards that are necessary to support regional products and services towards becoming internationally competitive. In the same breath, there must be adequate support for SMEs to enable them to access the services of accredited laboratories, where their products can be tested for compliance with specific standards requirements and assist in relevant training.

Globalization and Standards

Globalization has given rise to a new set of global standards while addressing a wide range of issues including labour conditions, health and safety norms, quality management procedures, and the environmental impact on production. Governments and enterprises in developing countries have found that while they are expected to comply with global standards, they make little or no contribution to the design of these standards nor effectuate changes to them.

Global standards are supposedly designed to improve efficiency in the world economy. Hence, the demand for standards has accelerated sharply due to the globalization of production and trade. Standards have been traditionally regarded as an important factor in smoothing trade relations and promoting efficient markets, by providing a set of common and widely understood benchmarks. By efficiently transmitting information, standards have reduced transaction costs. The ever-more complex interrelations that exist among producers, suppliers, retailers, and consumers across the world have accentuated the need for harmonization of norms and forms of codification.

Characteristics such as food safety, conditions for workers' safety and the credibility of products are embodied in standards that also give credence to the products that are produced. Without the strict implementation and enforcement of standards, it could be difficult for consumers to detect any compromise in the quality of the products they buy. Through standards, important information is transmitted from the business to the consumers. Because of certain common systematic approaches, local producers of goods are inherently integrated into the global value chain. Standards provide a common link among business enterprises and promote coordination between global production and distribution systems. Value chains have emerged as a powerful process in understanding how distinct functions that transform raw materials into traded end products are interlinked through complex arrangements among globally diverse actors.

Common standards such as technical norms, management standards and product codes promote compatibility among diverse actors within the value chain and create linkages among them. In addition to reducing transaction costs associated with value chain governance, compliance with global standards also lowers risks for various actors in the chain. Social and environmental concerns present unique challenges as well as opportunities.

For many years, standards have been used as a strategy to differentiate markets and create competitive niches. Compliance with global standards, especially on ethical, social and environmental concerns can be one important way to add value. The emergence of global standards has created new challenges for private and public governance both locally and globally.

The growing influence of standards in global markets tends to weaken national standards if not render them irrelevant in some cases. National standards are therefore increasingly required to comply with international norms. In so doing, the regulatory powers of national standards bodies are diminished. For example, in the food sector, HACCP as an international standard has become a mandatory requirement in most industrialised countries to ensure hygienic conditions and a consistently high level of product quality. In the US about 38 states have made HACCP mandatory. The European Union also introduced HACCP as a mandatory standard in 1993.

The application of HACCP requires public policies and the definition of the rule of governance in the utilisation of these HACCP processes and risk analysis. Many governments have integrated the standards into law, which reflects the seriousness of their responsibility to the public sector to increase transparency and thereby ensure that the health of the population is given priority. Strategically, some governments mandate domestic exporters like agribusiness firms targeting export markets, or newly de-regulated domestic markets to adopt HACCP to promote their competitive agenda.

Apart from public standards, there are important private initiatives that drive standards development. A case in point is the EUREP-GAP, the European Retailers Representative Group's standards on Good Agricultural Practices. From the late 1990s, EUREP-GAP has very rapidly gained wide acceptance in the European fresh produce retail sector with over 100 members, including prominent retailers and suppliers spread across Western Europe as well as countries supplying fresh produce to Western Europe, and "has authorized 20 certification bodies to carry out its audits in over 25 different countries." In the UK, for example, the standard has been adopted by the five leading supermarket chains that collectively account for 80% of total UK food retailing. The standard is in effect, an industry-wide response to formulating a single code that can offset the numerous country and firm-specific standards.

Notwithstanding the support and popularity of liberalised trade, the global economy is primarily governed by rules emanating from standards that govern trade. International standards have become dynamic in response to market conditions. Concerns regarding quality assurance, health and safety, as well as ethical, social and environmental aspects of production are now central to global trade. In developing countries, firms have come to realise that their capacity to compete internationally is often linked to their ability to comply with global standards. There is an apparent shift away from generic to sector-specific standards as in the case of food quality and safety standards. Many food safety standards have been defined by public interest groups or through public-private partnerships. Sanctions for non-compliance are either enforced by the market or applied by national and regional regulatory bodies.

Chapter 10
The Global Value Chain in Perspective

The value chain can be described as the full range of activities that firms and workers perform to bring a product from its conception to end-use and beyond. This includes activities such as research and development (R&D design, production, marketing, distribution and support to the final consumer. Value is created at every stage of the process and the total value delivered by the company is the value built up throughout the chain. The value chain concept was developed and popularised by Michael Porter in his 1980 book "Competitive Advantage of Nations." It was developed as a tool to assist enterprises to create value for their customers.

The value chain concept separates useful activities (which allow an enterprise to gain a competitive advantage) from wasteful activities (which hinder the company from getting a lead in the market). Focusing on value-creating activities could give a company many advantages. For example, the ability to charge higher prices; lower cost of manufacture; better brand image and faster response to threats or opportunities. It is a critical analytical tool that will help small enterprises to understand the system dynamics regarding waste, quality issues, efficiency and productivity and to identify critical areas for technical and policy interventions.

The value chain is also comprised of primary support activities including inbound logistics (getting the material into the facility and adding value by processing it); operations (all the processes within manufacturing); outbound logistics (distribution to the points of sale); marketing and sales (branding, promoting, and selling the product); and service (maintaining the functionality of the product, post-sales, etc.).

For the most part, a product is hardly consumed in its original form, but it becomes transformed, packaged, transported and marketed until it reaches the final customer.

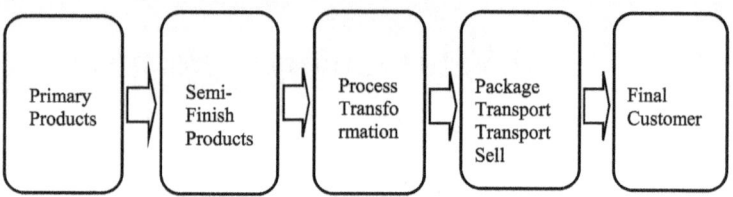

Figure 10.1: Typical Value Chain Flow Chart

The value chain sequence of activities should be first mapped to get a clear understanding of the main actors and relationships involved in the value chain. It would be important to include, for instance, those farmers who are the suppliers of the primary products and those who are involved in sorting, transportation, quality control and packaging, among other things. Although the various interest groups are known, their functions and relationships with other key players are often misunderstood. It is also important to note that the starting point of the value chain should be the end market where the buyers of the final product are. These are the people who determine the price, quality and quantity of a product or service. Cognisance should also be given to other providers of services, the external environment impacting the success of the enterprise, and linkages with other enterprises at different levels of the value chain.

The process of adding value could involve the upgrading of facilities and developing new products or services that will respond readily to market opportunities. As was established earlier, firms and industries need to innovate to add value to products or services and to make production and marketing processes more efficient. A value chain exercise is not prescriptive nor are there generic models that are being used. Effectively, there are useful procedural guidelines that will allow an organization to achieve its desired objectives. The chain analysis must ultimately focus on the constraints and opportunities identified during the exercise. A strategy to upgrade a value chain should be developed in consultation with and participation of all actors. The strategy must be systematically developed and there must be a commitment from the chain's actors who must take ownership of the process.

Challenges & Opportunities of the Global Value Chain

The global economy is increasingly structured around production and distribution systems that bring together diverse economic actors. The global value chain as it is called is a process involving the production of goods (as well as services) into finished products using inputs from various countries around the world. It is fast becoming a dominant feature in world trade and involves the participation of developed as well as developing countries. The global value chain becomes an operational process when buyers and sellers can engage in the trade of the necessary skills and materials that comprise the global production chain at competitive cost and quantity.

International production, trade and investments are often conducted within a globalized environment. Hence, different stages of the production process are located in different countries. This thrust of globalization particularly within industries has been facilitated by improvements in transportation and telecommunications technology. It is for this reason that companies restructure their activities by outsourcing to other countries while establishing dedicated operations in countries where there is a comparable favourable industrial climate. Within each segment of the value chain, the demand for value-added inputs has become very competitive. Today's supply chains are globally dispersed, and different activities are usually carried out in different parts of the world.

It is generally accepted that SMEs' productivity levels are much lower than those of larger firms since they are also known to have a smaller export base, which is a reflection of the strong orientation of the domestic market. Targeting the international market is an incentive for SMEs to augment their competitiveness so that they can tackle new and more complex markets. Entry into the global market gives firms access to the knowledge economy, innovation, network and market information which will position them to become world-class. SMEs in small states must seek to reduce the productivity gap between themselves and larger entities by value-added initiatives incorporating a greater level of innovation, technology and improved management systems. This will enable them to improve productivity to boost market share. It must be recognised that amidst enormous global challenges, the integration of SMEs into global value chains (GVC) directly or indirectly is an onerous task.

To overcome competitive shortcomings and barriers will require systemic changes that will allow them to adapt to a new paradigm within the global competitive market that will position them to capitalize on opportunities in the global value chain. For this to happen, it will be important that these small firms identify specific sectors within the global value chain in which they have the propensity and capacity to compete. They should also identify barriers and bottlenecks and devise strategies to address them with potential for SME participation by identifying markets and products.

In the global economy, countries participate in industrial development by leveraging their competitive advantages in assets. Developing countries often offer low labour cost and raw materials, while rich nations with highly skilled and experienced labour force place emphasis on R&D and product design in their quest for market dominance. Because of this approach, the level of the interrelationship among firms from different locations results in a greater impact on one another.

The expansion of the scope of GVCs will enable small developing states to improve their chances of becoming part of the value chain. Seeking foreign direct investment (FDI) will allow them to become beneficiaries of outsourced activities of production and services across international networks. The onset of global value chains in sectors such as agriculture, electronics and tourism can impact production, global trade and employment significantly.

Ideally, GVCs can provide a platform for small firms, customers and workers to build linkages through which they can develop business and working relationships, which can enhance the position of small firms to be integrated into the global economy. For low-income states, it can enable them to eliminate the need to develop complex production processes in-house and enable them to take advantage of more sophisticated capabilities within the external market that can complement their small domestic labour force.

Within the GVC, powerful retailers such as Walmart and Tesco as well as other successful companies with their branded merchandise like Nike and Reebok exert tremendous power and influence as buyers, by establishing and demanding what standards and protocols suppliers should meet as well as how resources are allocated and controlled. This explains the dominance of powerful actors that exist in the value chain. There are several benefits small enterprises can derive if they are so organised to be integrated and participate in GVC activities. Indeed, most SMEs from small developing states are constrained by geography, size, resources, access to markets and lack of development.

Fundamentally, these factors are determinants for participation in the GVC. Numerous policy factors can determine whether SMEs can participate and be competitive in the GVC particularly if the infrastructure are not in place to participate. It should be noted that GVCs are not exclusively designed for large multinational companies because of the unbundling of products and production processes.

The participation of small enterprises in a global economy is a vital undertaking, recognizing the obvious roles of small firms within the GVC. This may require a well-functioning business environment and vibrant entrepreneurial ecosystem to facilitate such participation. More importantly, small firms should not lose sight of the demand for technological change. Such change is contingent on competitive market forces and pressures; the innovative drive of firms; intrinsic flexible systems that will drive the firm's agility in responding to change and the access to resources that will facilitate that change. The entrepreneurial landscape and market dynamics may affect some small enterprises disproportionately and their ability to innovate and grow.

The ability to compete successfully in foreign markets and become part of the GVC is a formidable task for small firms that generally lack the capacity and resources. SMEs have found it extremely difficult to compete with larger companies along the entire line of activities in the global market if market failures that disproportionally affect SMEs are not addressed wholesomely. Notwithstanding those inherent drawbacks, it is a sensible strategy for any small firm to adopt. Such a goal will require specialised training and awareness regarding the benefits of internationalization of small firms as well as the entry modalities into these international markets. Small firms should seek to identify opportunities that are available in the global market and decide on what market niches they wish to pursue, thereby focusing on those activities in which they possess the expertise and then specializing in those tasks within the chain. They must, of necessity, build the relevant capacity in terms of the products or services they wish to offer and then establish the necessary linkages with transactional corporations with which they can collaborate.

On a matter of caution, SMEs' participation in the global value chain should not be such that they are relegated to the role of mere suppliers of parts and components to large corporations. If this happens, it can stymie their own innovation dynamics and expose them to a high degree of dependency on the principal manufacturers to whom they supply. A careful assessment is therefore required to ensure that value chain integration does not run counter to SME development strategies focusing on the promotion of direct exporting capabilities. As discussed earlier, small firms can address many of the disadvantages they face by adopting new industrial policies that are outward-looking in design. They should therefore seek to develop linkages with larger firms from which they can benefit. SMEs with higher growth potential that are integrated into the GVC often grow to become leading firms by undertaking higher value added tasks. Other policy measures include:

i. Opening borders and attracting investment can help jump-start entry in GVCs.
ii. Countries will derive the greatest benefit by maximizing the absorption potential of the domestic economy and by strengthening linkages with GVCs.

iii. Many diverse policy areas affect the success of GVCs. They include, among others, trade policy, logistics and trade facilitation, regulation of business services, investment, business taxation, innovation, industrial development, conformity to international standards and the wider business environment fostering entrepreneurship.
iv. Countries should identify measures that will complement their GVC strategies. These could include several developmental thematic priorities including investment in education and vocational training, building environmental consciousness, urbanization shift, developing ICT infrastructure and labour market mobility.

Functioning within the Value Chain Systems

A supply chain is critical to the operation of every small business enterprise. No market can function effectively without them because they are critical to the survival of any business. A supply chain may be defined as an integrated process wherein several business entities (suppliers, manufacturers, distributors and retailers) work together to (1) acquire raw materials, (2) convert these raw materials into specified final products, and (3) deliver these final products to retailers. This chain is traditionally characterized by a forward flow of materials and a backward flow of information.

Competitiveness in the global economy is highly dependent on an effective supply chain management system. In the current global scenario, many manufacturing firms are not necessarily engaged in the manufacturing of some projects. They outsource a significant portion of their production activities to smaller international companies, many of which are located in developing states. This decision allows them to maximise their productivity gains and augment their competitiveness. Although some components of their products are manufactured elsewhere, they however, retain the product brand name. The brand value or the commercial value which is derived from consumer perception of the product is determined by the contracting firms' performance in the manufacture of that component of the product. It is in the interest of the company to develop the relevant capabilities such as quality systems, logistics providers and other support infrastructure to support these smaller contracting firms.

Firms should also be concerned about the cultural, social and environmental conditions in overseas plants and should make the necessary provisions to address them. Failure to do so may place the brand image and sales at risk. Effective brand management is an integrated process in which all supply chain partners cooperate in the coordination of production schedules and logistic activities so that there is mutual benefit for all. The level of cooperation must transcend the mutual sharing of information vital technology and expertise, which will lead to increased sales due to lower costs, improved quality performance and higher productivity.

The supply chain is about establishing forward linkages with customers and backward linkages with suppliers. At the domestic level, backward linkages relate to the sourcing of parts, raw materials, components and services for SMEs. It is one of the most important functions in the growth process of SMEs. The focus is not necessarily on an adumbration of how the supply chain functions within the global value chain but on the role of a supply chain in assisting small enterprises to become competitive. The modality of functioning within the global value chain involves a complexity of issues involving suppliers, distributors, marketing, technology partners and the like, which is beyond the scope of this work.

Procurement of goods and services typically represents the largest single category of spending for SMEs, ranging from 50% to 80% of revenues in most manufacturing organizations. Thus, improving quality and productivity through effective supply management practices is critical to SMEs. Suppliers are as important to the organization as any other interest group. Sometimes the relationship between suppliers and the firm can be controversial and therefore needs to be managed. The primary goal of the enterprise is based on cost, quality and timely delivery. Purchasing supplies is considered an integral part of the overall goal of a firm's strategic goal. Hence, the firm must maintain long-term supplier relationships.

It should be emphasised that suppliers should be integrated into the buying organization's core processes rather than simply keeping them as vendors at arm's length to gain a lasting competitive advantage through a sourcing arrangement. At this level, supplier-customer relationships are not viewed as just buying or selling products or services, instead, they are considered extensions of the buying organization. The main objectives of an effective and efficient supply chain are included in the JIT (Just in Time) concept discussed earlier.

The concept entails the exigency to deliver the right product at the right time, in the right amount and at the right place. To meet these demands, goal oriented activities should be given priority to ensure that all processes of the supply chain function efficiently.

It is vitally important for all enterprises irrespective of size to develop a supply chain management system aimed at ensuring both short term and long term effectiveness and efficiency of the supply chain processes. The traditional approach to the management of small and medium-sized enterprises is associated with overstocking and providing excessive capacity as a means of protection against demand variability. Sometimes there is a process of determining when to replenish dwindling stocks and the firm often faces the challenge of shortages.

As a result of the possibility of rapid and major changes in the marketplace, this approach presents a great risk with potentially adverse effects. In response to the demand variability, all the elements including demand management, planning, procurement, warehousing, production, transport and distribution that constitute the supply chain must be effectively streamlined to ensure there is optimal delivery of service.

The Agriculture Sector and its Linkages to the Supply Chain

As the population of many developing countries grows, there is an obvious demand for more and better-quality food. Agricultural resources such as land and water are stretched thin and the yield per acre is diminishing rapidly. The sector continues to be impacted by the deleterious effects of climate change in the form of reduced rainfall and severe weather conditions such as droughts, hurricanes and floods that affect crops and livestock. Because of these ongoing issues, agriculture requires a cadre of farmers with a new mindset supported by technological innovation.

The vision of developing a vibrant agriculture sector embodies the fulfillment of food security goals through the sustainable use of a nation's agricultural resources. This is intended to meet the needs of a growing population utilizing the production of wholesome and nutritious foods based on the sustainable use of the natural resource endowment of the state. The sector's development should be catalysed based on innovation, science and technology including good farm management practices that would enhance competitiveness in all segments of the value chain. The growth of the sector is dependent on domestic needs and market expansion including regional and extra-regional markets, through value-added enterprises. This will enhance the competitiveness of agricultural enterprises in a manner that will increase income for farmers and contribute to economic development.

The agricultural thrust in many developing states, for example, continues to be oriented towards primary production and mono-cropping. Notwithstanding years of unfortunate experiences and poor policy choices, first with sugar cane, then cocoa, citrus and more recently bananas, much has been promised about the transformation of the sector by adding value to primary products in agro-processing. After many years of anticipation of unprecedented growth within the sector, the investment necessary to transform it has not materialized. Allocations from national budgets for research and development to kick start the process have been quite inadequate. For example, following the closure of the Produce Chemist Laboratories (PCL) in most of the small states of the Caribbean in the 1970s, there have not been appropriate replacement facilities to drive research and development at the national level. At the same time, the other countries in the region have developed Bureaus of Standards with the main purpose of being regulatory bodies and not product development. In some instances, donor funding has been used for agronomic and post-harvest research and not for product development activities.

For many years, agricultural development in some regions was beset by a disjointed approach among three related components viz. health, nutrition and manufacturing. Realistically, agricultural production is geared towards the export market. Except for rice and sugar, there are few linkages between the agricultural products that are produced and what has become part of the diet of the local population. In most cases, the approach to meeting the needs of the local population is primarily based on any surplus from what is being produced, which is in stark contrast to developed countries in which the surplus of what is produced is exported. In small developing economies, agriculture is seen as a foreign exchange earner mainly through the export of primary produce.

In developed countries on the other hand, substantial benefits from agriculture are derived from the transformation of primary products to value-added processed products that are exported, much of which form part of the cuisine of those populations. For example, regional hoteliers are often reluctant to introduce Caribbean cuisine as part of the menu made from indigenous tropical products as part of the visitors' Caribbean experience.

Agriculture should be perceived and approached as a business in all its forms. It is generally understood that agricultural development in small island states is replete with several binding constraints at every level of the supply chain, including the value chain activities. Sustainable agriculture development, therefore, requires measures that would address the binding constraints identified throughout the value chain, from production at the farm level to promotion and distribution to the final consumer. The transformation of the agriculture sector will depend on the involvement of the private sector, which is reputed to be averse to taking risks associated with agro-processing, preferring instead to trade in primary products.

To identify the challenges that beset the agriculture industry, an understanding of the nature of the industry and the key operatives is a fundamental requirement. In most small economies, small farmers constitute over 80% of the entrepreneurs who are involved in primary production. Many of these farmers are not well educated, they hail from poor rural backgrounds and depend on agriculture for a livelihood. The industry continues to be plagued by the reluctance to use advanced information technology. Many farmers (particularly the older ones) do not follow instructions, or they sometimes try to get more from the products used as inputs than is otherwise recommended. This is especially damaging regarding the use of pesticides, where if too little is used the pests become resistant quickly and the crops are poisoned if too much is used.

Generally, there is no established medium to facilitate farmers' interaction that will seek to promote linkages and synergy among interest groups through which information regarding market intelligence, quality concerns, opportunities for trade, new technology including farming practices and activities within the supply chain. On the one hand, producers and marketers require intelligence on available markets, buyers' preferences, product specifications, prices and delivery schedules.

Regarding the issues of the specification of raw materials, several factors should be considered in delineating the quality characteristics of raw materials that meet the requirements for processing. Most of the negative quality issues threatening the quality integrity of the final product have their origin in the quality of the raw materials that enter the facility. Quality attributes such as appearance, size and the extent of mechanical damage are important attributes to consumer appeal. The hygiene of the finished product must also be of major concern. Hence, the raw material should be free of contaminants, pests and disease. The freshness and maturity of primary products and their physical state are quality considerations that should be taken seriously. A strong and effective communication and monitoring system within the supply chain can avoid many of the unsavoury issues that may likely occur in the manufacturing of a product.

Along the supply chain, beginning at the farm gate, there needs to be competent management of the system and its resources to achieve high levels of operational production efficiency. The application of technology in achieving stipulated quality standards such as shelf life, packaging and labelling is of paramount importance. Achieving operational efficiency throughout the value chain, from production to the market is an ongoing process that starts at the farm level where competitiveness in production requires producers to become cost competitive as well as to supply quality products. At the farm level, technology is required to ensure efficiency in production including irrigation, high yielding varieties and effective crop protection.

Small farmers operate under several serious constraints that could determine their survival in the industry. The application of technology embodied in Good Agricultural Practices (GAP) as part of modern farming techniques may to a large extent be resisted by those farmers who have not been formally trained and are not technology savvy. Most of these farmers lack the requisite resources to conduct any substantial farming, and many of them do not have the necessary collateral to secure loans from financial institutions for investment in farm equipment and input. Historically, SMEs, particularly agri-based enterprises began their business operations by focusing primarily on the production of primary products. Over time, many of them have mushroomed into producers and suppliers of finished and semi-finished manufacturing products or they have become service-oriented organizations in the distribution industry.

Value Chain Challenges and Impact on Key Sectors

Traditionally, many small ACP states have been dependent on agriculture as a major foreign exchange earner because it has contributed significantly to the GDP of these states. Over the last couple of decades, agriculture has been in a state of decline in many small states. The sector is generally comprised of traditional export commodities including fruits, vegetables and speciality crops. The statistics do not generally include the agro-processing subsector, which is included in statistics on the manufacturing sector and therefore understates the impact of agriculture. The production of traditional export commodities has declined due to increased competition and the inability of many countries to be internationally cost-competitive.

Finding solutions to arrest the rapid depletion of the world's resources is an ongoing exercise for policymakers. The transfer of technology can ideally be used as a tool that can be applied to sectors like manufacturing and agriculture with resounding success. Traditionally, agriculture has primarily been labour intensive with minimal technological infusion. Although the agriculture sector in some countries has been modernized using modern technology, there are areas within the sector that need more modern technological intervention. In recent years, the marked decline of the banana industry in many small island states for example, has resulted in a large proportion of the agricultural lands being converted to residential sites to provide housing for the growing populations. The remaining lands are woefully overworked and have been seriously depleted of essential plant nutrients. Smart agriculture, using technology seems to be the only tool to address the issue of land shortage and low productivity.

Most enterprises procure the raw materials offered to them by small suppliers, but do not actively encourage production through supply chain linkages such as contract farming, out-grower arrangements, or product collection networks. Through effective participation in supply chains, rural producers and agribusinesses will increase household incomes, sales volumes and create more jobs. Agribusiness enterprises have been reputed for variations in standards. Failure to achieve consistency in the standards required to compete effectively has hindered their opportunity to enter foreign markets. Common weaknesses that can be attributed to such weaknesses include (i) lack of technical skills for effective production management, establishment and maintenance of supply chains and food safety standards accreditation; (ii) inadequate export planning and marketing; and (iii) weak business and financial management.

There is a perennial problem that plagues many small enterprises which relates to how raw materials are being sourced. Most of these states are endowed with an abundance of tropical products that form part of the agriculture export trade. The domestic agro and agri industries are supplied with agricultural products that are produced seasonally. To a large extent, production is unplanned, because most of these crops grow on very small family holdings around the islands and no specific care and attention is given to harvesting the fruits in a structured manner. There is generally a high degree of waste due to spoilage when the fruits are ripe and drop on the ground without any organised means of gathering them for future use.

For seasonal fruits, there is a copious supply during the season of high production and farmers are arbitrarily contacted by agro-processing enterprises to supply these fruits. However, during the off-season, there is a noticeable shortage of those fruits as well as the products that are derived from them. On the other hand, agri and agro-processors sometimes enter into loose arrangements with farmers to supply selected crops such as peanuts, sorrel, ginger and pepper. Those crops are often grown in small quantities by small farmers for fear that they may oversupply the market and thereby stand to lose. Sometimes suppliers are not even paid for their crops. When this situation occurs, there is a reasonable reluctance to provide further supply to the enterprises.

A well-developed and organised supply chain system would correct those deficiencies and ensure that there is a relationship between the business enterprise and the suppliers. Such relationships should be built on trust which will be engendered by binding contracts to supply raw materials, components and equipment to business enterprises. Moreover, since many of the poorer farmers are engaged in subsistence farming, assistance should be given to them to encourage them to expand their business by enabling them to secure credit for the purchase of farm equipment and supplies.

There is also an issue regarding the quality of raw materials being supplied which is of serious concern to business enterprises. For the most part, farmers lack the knowledge regarding quality and standards requirements as well as modern agricultural practices that will allow them to produce crops that will meet specific standards. The onus is on the manufacturing enterprise to provide the requisite and systemic training in modern agricultural practices aimed at producing products of peculiar variety that will give increased yield and are more resistant to specific crop diseases and drought.

Challenges in harvesting and logistics are ongoing issues that need to be addressed to prevent crop loss through spoilage. Selection, storage and transportation are critical to effective harvesting. Crops that are transported over long distances may require specialised mode of transportation to transport them from the farm to the factory, which in some cases may be located in remote areas. The challenge of the seasonality of fruits and the shortage of products that often emanate from them can be best addressed by investing in specialised equipment that would facilitate long term storage. Apart from specialised storage equipment, enterprises may be required to invest in additional storage facilities retrofitted with cold storage and controlled temperature mechanisms.

A typical example of the foregoing situation is in Samoa's agribusiness industry, which has been plagued by this drawback for many years. They have been unable to procure sufficient raw materials of consistently good quality. As a result, producers of virgin coconut oil were unable to fill existing export orders. The country also lost significant revenue because of severe shortages of appropriate fresh varieties of taro for processing into snack foods for the export market.

In light of the foregoing setbacks, attempts should be made to train farmers and revolutionise their thinking to embrace modern technology. The appropriate infrastructure including access to technological innovation and a farmer support network should be developed to establish a competitive and sustainable fresh produce business. This is particularly important to address the challenges at the post-farm level of the value chain including activities such as raw material collection, handling, packing/processing, storage, transportation, distribution and sales. Achieving high levels of operational efficiency requires economies of scale in the various post farm activities. The application of technology is also central to achieving value-added, through the development of products to meet market demand.

Some proponents argue that agricultural development policies should focus primarily on larger commercial farmers. It is widely believed that small scale farming is generally subsistence in nature and does not provide income for sustainable livelihood nor is there a surplus supply of food for the domestic sector. Subsistence farming should be encouraged in situations where farming is not the principal economic activity and the goal of reducing the food bill and providing quality food is a priority. It is within this context that agricultural development policies should be all- encompassing and the programmes that support those policies should also provide for the needs of the entire sector. On the contrary, agricultural development is not only about feeding the urban population, providing employment opportunities, and generating profits. There must be an element of sustainability to the food and nutrition security of the nation.

The perennial challenges facing the sector must be resolved before developing countries can take advantage of the opportunities that are available through participation in the GVC. Issues including climate change, continuous degradation of natural resources, low technology utilization, and many others are constraints that plague the Agricultural sector. These challenges present a greater burden on the rural farmers, particularly those with small holdings. The small-scale farmers rely primarily on family members as a means of cutting costs. Family members who provide labour on farms are regarded as mere helpers and not as paid employees. The perception is that they are simply offering labour to compensate for the meals and lodging that is provided to them at home to which they do not contribute financially. Refusal to work on the farm creates a litany of domestic problems including abuse.

There needs to be a shift in culture and thinking to view agriculture in the appropriate context, that is, as a business venture that requires a similar approach to any other commercial business that requires record keeping, resource allocation including labour and wages, and the like. As was discussed earlier, the infusion and science and technology can resolve many of the existential challenges facing the sector.

The amount of land that is available remains constant. The statistics show that the populations in many small island developing states are gradually declining. Hence, it is not anticipated that the land-use pattern will change in the short term, where large proportions of arable lands will be converted to residential use because of the demand for housing. The global value chain holds tremendous opportunities for agriculture. The actual impact of the global value chain on economic performance is not known because the data is not available. The Global Value Chain should be perceived as the catalyst to foster the structural transformation from a subsistence-oriented and farm-centred system to one with a commercial agenda.

The expansion of the agricultural sector is highly dependent on the integration of the regional market for trade in agricultural produce. Several factors combine to restrict intra-regional trade in agricultural products. These factors can be summarized as follows:

- Failure to implement harmonized Sanitary/Phyto-Sanitary protocols for agricultural commodities;
- Limited and costly intra-regional transport;
- Absence of a Market Intelligence System (one that could assist in identifying trade opportunities and improve overall market operational efficiency;
- Limited capacity to facilitate intra-regional trade in agricultural produce e.g., cold storage, packing houses, shipping and handling facilities;
- Failure to implement a universal quality assurance standard e.g., global Good Agricultural Practices (GAP).

As in the case of manufactured products, the agricultural sector is also required to comply with global market requirements that promote competitiveness for entry into several regional and global markets. For example, regarding the banana trade, bananas from ACP states were required to comply with liberalised trading conditions imposed by the WTO following the removal of the protective regime that previously existed. Intermediaries in the supply chain are responding to regulators and consumers by demanding that suppliers provide high-quality products that are safe and wholesome. Suppliers are also required to provide evidence of traceability to the source of production.

Chapter 11
A Strategic Framework to Support SMEs

In attempting to find ways to improve the competitiveness of enterprises, it is necessary to devise a systemic and logical set of procedures that involve a sequence of activities that are considered prerequisites in making decisions and taking actions towards strengthening a firm's position through progressive and incremental improvements. The set of procedures depicted in Fig.11.1 is indicative of a process that will enable an organization to examine its options, activities and resources and combine them in a manner that will provide synergy and complementarity. This approach will enable the firm to respond expeditiously to its competitors as well as to effect the necessary changes dictated by the market. This exercise will preempt various options that will stimulate progress towards becoming a world-class organization.

Strategic Continuous Improvement

The Strategic Continuous Improvement Process (SCIP) is intended to be a model process, which was developed as a generic tool to be used by any organization. It intends to embark on a systemic approach to enable an organization to improve its competitiveness and attain world-class status. It was developed to avoid the pitfalls that many organizations encounter and to which they arbitrarily and fortuitously commit enormous resources. In many cases resources geared towards improving aspects of a firm's processes are expended in a piecemeal manner without understanding, paying attention to, or adhering to sound protocols on issues of good manufacturing practices and how they are interrelated. This process in question intends to bring discipline and order to improve competitiveness within the organization in a coherent manner. It also encapsulates the main elements of the organization's strategic assessment, planning and implementation processes as they relate to the production of goods and services. It further demonstrates how various elements of the production system can be integrated synergistically.

The activities depicted in the SCIP process should be followed as proposed, providing that an organization had already conducted some of the initial activities. Those activities that are considered fundamental to the organization should not be omitted or overlooked in preference for more advanced activities, since the basis on which to establish strategic investment would be lacking. Any attempt to do so will seriously undermine the integrity and usefulness of the process. Tools such as SCIM Process could be used for the evaluation and improvement of internal operations. These tools can be used to assist enterprises in conducting a methodological and effective evaluation of the critical areas of operations to determine where improvements are required.

Assessment Phase

This phase begins with an assessment of the company's current competitive position in the market. Each firm should decide on a set of criteria that are important to its competitiveness. It emphasized that competitiveness should be directly related to competitors' strategic strengths and prevailing market requirements. Typical criteria on which competition is based include product, price, quality, service, flexibility to respond to change, product and process technology, efficiency and productivity and design. An organization should therefore assess these factors against those of its competitors.

The firm should also conduct a review of its corporate strengths, weaknesses and business objectives to determine their relevance to any new corporate undertaking. Organizational strengths must be sustained over time, which should be able to place the firm in a comparatively advantageous position. On the other hand, weaknesses should not be debilitating to the firm's ability to respond to change and thereby compete. Moreover, the firm should be able to translate weaknesses into opportunities.

Weaknesses such as inefficiency, poor quality, and low productivity are symptoms of internal problems that management should investigate and correct. The corporate objectives could be developed simultaneously within the framework of the strengths of the firm.

The Planning Phase

The firm should identify key performance areas. Performance is a measurement of how efficient, effective and adaptable an organization is. Performance areas are manufacturing and business activities that are critical to the organization in achieving its objectives. Performance areas include product performance, service quality, product features, manufacturing technology and post-sales services. Having identified performance areas, performance targets should then be established. Those targets should relate to specific measurements from which achievable goals can be measured. Those measurements should be efficient and effective.

Efficiency is the extent to which resources are maximised and waste is eliminated. Productivity is a measure of efficiency and is measured by factors including cycle time, waiting time per unit, resources expended per unit of output and processing time. Effective measurement is a measure of how well the output of the process or sub-process meets the expectations of its customers. Effective measurements include accuracy, adaptability, appearance, costs, dependability, performance, reliability, serviceability, usability and timeliness. It is important to note that quality is a factor which is determined by effective measurement. The following is a list of typical measurements that are required in routine production activities.

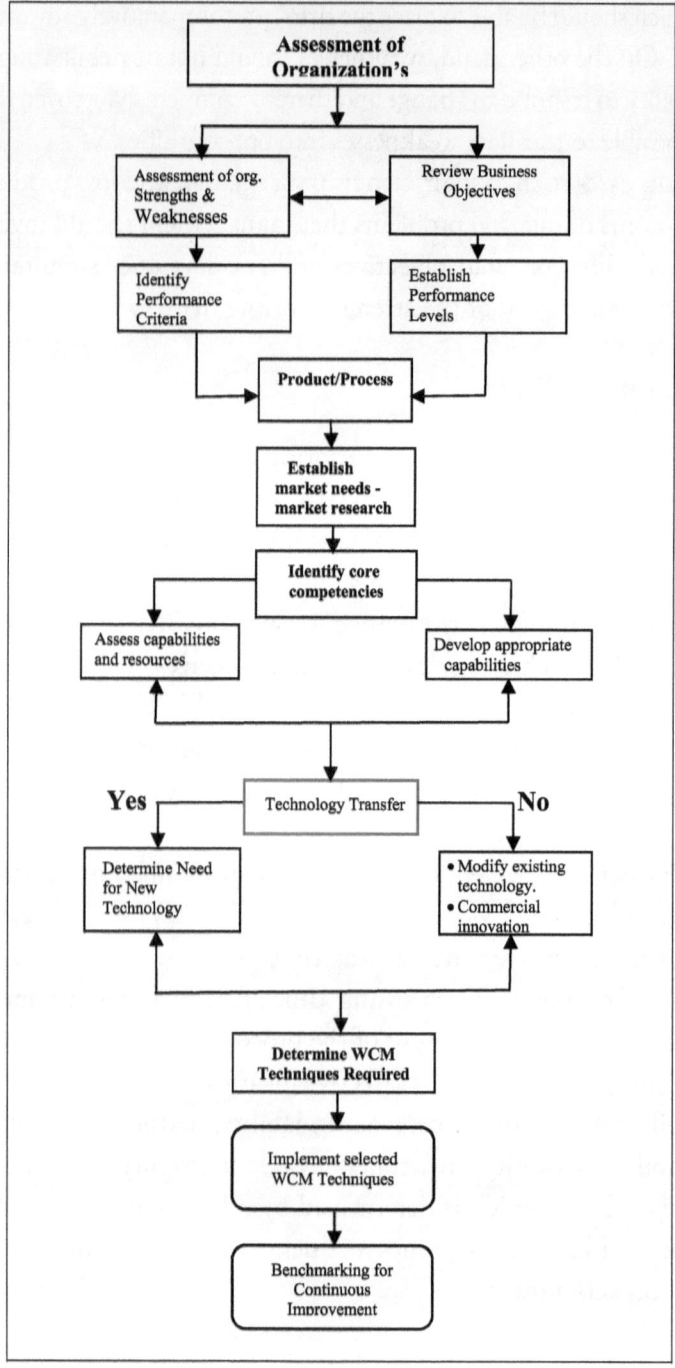

Fig. 11.1 Strategic Continuous Improvement Process

Having identified the main performance areas and established performance targets, a firm should evaluate its products and processes to ascertain if it can perform to the required levels. This exercise will allow the organization to make the necessary adjustments to processes and to examine the relevance and performance of its products. The organization will then be able to begin the planning process to meet future targets. Firms should update their information database to respond to the demands of the current state of the market vis á vis consumer needs and preferences. This can be done through extensive market research, which should be an ongoing process. This exercise can be facilitated through the development of a database on market intelligence to enable them to respond to market changes and challenges.

It is in the interest of firms to identify core competencies that are required to meet various organizational needs. Core competencies are skills that an organization needs to adapt to meet current and future needs. Such skills could include technical knowledge that would result in the development of capacities in storage technology, miniaturization, packaging and reduced lead time. An assessment should also be conducted on the capabilities and resources to ascertain whether these required competencies are available, adequate, and relevant. Deficiencies and shortfalls in capacity and resources should be addressed speedily if an organization were to acquire or develop the appropriate resources and build the relevant capacity it needs.

Implementation Phase

In the process of developing the appropriate technology, an important question that has to be addressed is whether the firm has the technological capability to grow and flourish in the market. This would require a decision-making scenario in which the question of technology transfer may arise. Should the firm pursue technology transfer? If the answer is yes, it may have to determine what new technology is required, the cost of such technology and how it is going to adapt to new circumstances. It may not be necessary to acquire new technology in which case the firm could modify those that exist. In some situations, it may or may not be appropriate to do so. Should it use commercial innovation in its marketing and promotional efforts to reposition the products in the market, thereby giving them a new appeal? Does it embark on a radical technological approach to new product development?

Some world-class manufacturing techniques may not be applicable to enterprises in small states as the scale of production activities and complexity may not require sophisticated systems. Firms should decide on which techniques apply to their own circumstances and are vital to their ability to compete with the best in the business. The emphasis should be placed on techniques that will drive quality, employee involvement, productive maintenance and so forth.

When a firm has decided on the appropriate selected WCM techniques, it must also take measures to implement them. This implementation can be done in its entirety or on a phased basis. For the most part, it is necessary to develop an implementation plan which would involve keeping all employees abreast of various developments within the organization, thereby minimizing any disruptions in the manufacturing process.

After organizations have invested significant resources in improving efficiency and productivity, they need to benchmark their performance against world-class competitors. This process involves an evaluation of various processes, products, and services against the best performing enterprises. This comparative analysis serves to recognise gaps between its current performance levels and world-class performance as characterized by those companies selected. A generic benchmarking software will be a useful tool to assist in this exercise. The model intends to provide a guide to firms in their quest for efficiency and competitiveness.

The SCIP process in question is not an end in itself, but a useful tool that may be likened to a compass, which will guide the activities of a firm. There are checks and balances inherent in the process to enable it to maintain its essence and reliability. In implementing SCIP, a firm may not necessarily begin at the assessment stage.

The exercise should begin at the stage where the firm has identified weaknesses in terms of inefficiency, waste and low production, among others. A useful exercise would be to develop several evaluation criteria that are critical to an organization. This list of criteria could be important and can be used by any organization to evaluate various processes.

Productivity and Productivity Improvement

The importance of productivity is universally recognised. Productivity can be described as the output generated by a production system (a good or a service) from the input provided to create that output. The word productivity has been used widely and with varying meanings. For this reason, the calculation of productivity has long been associated with controversy. Quite often, efficiency is confused with labour efficiency. It should be recognised that factors such as an increase in the cost of energy, raw materials cost, and the technology employed would impact the increase in labour productivity and should therefore be considered in its determination.

Generally, productivity relates to how efficiently factors of production (land, labour and capital) are utilised in the production of goods and services. It is a measure of performance for the production activity and refers to the amount of output produced per unit of input used in production, which is generally expressed by the following equation: Productivity = Output / Input. This formula is referred to as partial factor productivity because it constitutes the ratio of total output to a single input. It is widely used because the data is easily available. The formula is based on the principle that an (output) is derived from an (Input) of (labour hours, materials, capital and equipment); whereas sales or the number of goods produced are common output units. Some of the most common productivity measurements are total employee labour productivity, individual employee labour productivity and sales productivity. To increase productivity, either more output is produced with the same amount of inputs (labour, equipment, and tools) or fewer inputs are required to produce the same level of output.

One of the major challenges in measuring productivity in real-time is that it is subject to changes in the data source. This formula shows which change in input and output should be measured inclusive of both quantitative and qualitative changes. The levels of productivity will be determined by the competence of management and how resources are allocated. Economic growth is generally predicated on increased productivity. At the operational level increased productivity reduces the cost of production and enhances competitiveness through better quality goods. When there is a downturn in economic activity, it is reflected in a decrease in the level of productivity. Although productivity may have different meanings to different people, it nonetheless has relevance to the efficiency of the production process and on the quantity and quality of goods produced in relation to the quantity of goods used to produce them.

The productivity formula should consider measures to evaluate the performance of productivity improvement programmes. Productivity and efficiency reflect overall performance. Performance measurement is a critical aspect of the general management process and must therefore be considered as a viable tool to evaluate the performance or the level of operations in organizations from which adequate action could take place. Managers generally use productivity measurements to determine which departments, plants or workers are most efficient and how to maximize the company's resources to attain an optimum level of production and fetch higher sales and revenue. These measurement programmes should be such that they are justifiable and reliable for use in continuous monitoring and control of organizational performance. Productivity improvements engender profitability and are usually the main driver for managers whose main goal is to maximize profit. It is also a function of the utilization of financial resources.

Apart from the basic formula described above which is used for the calculation of productivity, there are different ways to measure productivity across different activities. Despite the various productivity measurement methods that are available, many of them are not widely used because of some complications involved. Productivity improvement involves more than a relationship between input and output. In an era of scientific management, there is a plethora of management tools as well as a wide range of technology applications in machinery and production systems that relate to productivity performance that are transforming the production landscape.

There is a stark dichotomy between productivity and quality improvement. However, in the new productivity paradigm, the thinking is, that firms should aim to achieve better productivity results and at the same time achieve improved quality products instead of the traditional production approach, where the focus is on productivity improvement. As firms strive to improve productivity and quality simultaneously, the focus should be to target competitive priorities relating to cost, quality, flexibility and delivery. They pursue a systemic approach that will require radical changes in the norms and culture of the organization. For example, the Six Sigma quality programmmes and lean production systems with their toolkits are notable programmes that have been successfully used by many firms over the years. This new approach offers tremendous benefits including improvement in the firm's flexibility, cost reduction, shorter lead times, elimination of wastes of varying kinds and its overall global competitiveness.

Improving efficiency in critical areas such as industry, agriculture and services will require the incorporation of technology; stimulating innovation, competition and quality; eliminating market distortions that impede the growth of the most productive firms; removing trade barriers and advancing international agreements that facilitate exchange. If the appropriate policies are applied, the potential of most small states can be maximised to result in economic growth.

The application of increased automation and the widespread use of information technology are not the only approaches to productivity improvement. Managers have at their disposal several other approaches including processes and work methods that they can use to improve productivity. These methods can be particularly applicable where it is difficult to apply automation as in the case of the services industry.

In the absence of mathematical approaches to calculating productivity, several techniques are being used to assess productivity improvement in the workplace from which strengths and weaknesses are identified for improvement. These programmes could fall into two categories viz. (i) The technical approach and economic analysis (ii) Human approach – behavioural methods. The former includes a combination of method study (motion study) and work measurements.

These techniques are used to examine people's work and highlight the factors that affect efficiency. Work-study is conducted to increase output from a given quantity of resources. The process involves systematic analyses of existing operations, processes, and methods.

Method study helps to eliminate unnecessary movement while work measurement helps in investigating, reducing and eliminating ineffective time when useful work is being performed. Work simplification, Pareto Analysis and Just-in-Time are behavioural methods that are also used in improving productivity. The responses will indicate possible areas that need attention and could be targeted for benchmarking.

The proposed framework which has been provided, if implemented in part or wholesomely, can assist small and micro firms to address several rudimentary challenges. Such challenges have stymied their growth by weakening their management capability, infrastructure and logistical orientation as well as their technical and technological proficiency for many years. No one approach can fix a firm's shortcomings. Based on some of the approaches and tools adumbrated earlier, a firm should decide which tools, technology or techniques are best suited to address its challenges. For change to be meaningful, it cannot be a business-as-usual approach.

Technology transfer in conjunction with selected world class manufacturing techniques if used strategically in SMEs as part of the Strategic Continuous Improvement Process (SCIP) can yield significant results and transform organizations to become world-class. Commercial innovation responses can also transform the fortunes of small and medium-sized firms.

Most of these firms are already established in the market, and their products have attained the maturity stage of the product life cycle. At this juncture, the SCIP may direct management to alternative solutions to which they can revert. Among the alternatives are the choice of world-class manufacturing techniques and other technological interventions and innovations that they can consider in relation to the firm's circumstances and challenges.

Benchmarking within the SCIM Process

Benchmarking is also an integral aspect of the SCIM Process. It is a useful tool for comparing a firm's activities to baseline industry standards and identifying areas for improvement. There are some exogenous factors such as fiscal and regulatory measures, environmental protection policies and those emanating from free trade to which firms must radically respond but are not addressed by the process. Although these factors are not immediately and intimately related to a firm's operations, they, however, can influence the level of functionality regarding the level of investment and profitability.

Experiences in life are meant to teach valuable lessons. An organization must likewise learn from its mistakes and those of others. This is done by devising a comprehensive programme in which it can benchmark its activities against those of its competitors to identify gaps between itself and world-class competitors. It should then devise a plan of action aimed at minimizing those gaps and thereby keep improving continuously to ensure sustainable competitiveness.

It was stated earlier that the essence of benchmarking is to look at and learn from others by comparing oneself with them. This involves the entire management team searching for the best practices outside of the company and examining those management and manufacturing techniques and strategies that will assist them in closing those gaps and adapting those techniques that are appropriate. Supporting evidence abounds that world-class companies have strong management structures, good and incisive, proactive and change-conscious management, which will employ the appropriate technologies, encourage employee involvement and build strategic relationships with suppliers. This is a learning process, which will inform the process of continuous improvement towards attaining world-class status.

One of the features of benchmarking is to examine and compare processes, management techniques and strategies that produce the results. Notwithstanding the heterogeneity of some manufacturing orientations, the aim is to identify best practices wherever they are found. For this exercise, Generic Benchmarking will be most appropriate where comparisons can be made across functions or different industries. Success in manufacturing cannot be attained without addressing wider national issues that will impact the growth of enterprises. Among them are industrial and fiscal policies, enabling legislation, adequate and appropriate infrastructure for industrial development and foreign policy such as those that are influenced by trading arrangements within trading blocs.

Although many relatively larger firms in some small developing states are facing the onerous task of surviving in the global market, many of them are, however, well established in their specific local markets and to a limited extent, regional markets. The emphasis will be for them to adopt appropriate techniques that would allow them to become efficient so that they can compete with the best in the business.

Organizational Assessment

The Strategic Continuous Improvement Process (SCIP) that was explained earlier was developed as an approach to assist SMEs to rationalize their operating system to determine what aspects of the model can be applied to their assessment and improvement needs. It is important to keep in mind that no enterprise is obligated to change its traditional practices. Their own experiences in respect of their position in the global market and the success (or lack of it) associated with those experiences could be an impetus for change.

Operationally, some organizations are not at the level of readiness to pay any serious attention to the notion of adopting world-class manufacturing techniques and the infusion of innovation, partially or in its entirety. The SCIP can be used as a tool for internal evaluation and improvement, which can then raise the consciousness of the need for change to become more competitive. It was also developed to inform organizations about their strengths and weaknesses, resources requirement and the relevance of their corporate strategies so that management can respond with appropriate measures to strengthen specific areas.

The success of the model and other initiatives will be largely dependent on the seriousness and commitment of the organization. For example, if the shareholders of an organization are fully committed to its advancement, the whole organizational structure may be revamped, thereby engendering a resurgence in the form of an injection of capital to revitalize the operations. This will inspire futuristic thinking in terms of new initiatives to expand product lines and break into new markets.

Some organizations may have implemented quasi forms of world-class manufacturing techniques but have not fully implemented them because they lack the expertise and experience to do so. In some cases, there may be a lack of knowledge about new technological approaches and innovation that could transform the organization, but some firms may lack the resources to engage professional consultancy services. Some firms may have attempted to conduct benchmarking exercises but may have failed to accomplish the desired objectives.

When these circumstances arise, it is common for senior management to become defensive when questions are raised about the modus operandi of the firm's current operations, particularly if they are antiquated and obsolete. This could lead to an attitude in which management is not too receptive to new ideas and is therefore likely to continue in its traditional ways. Conversely, some managers may give the assurance that the relevant strategic actions will be implemented to ensure the future success of the organization.

It can be concluded that firms in small states need to pay more attention to what is happening at the global level, which they should target as their ultimate operating environment. They need to be sensitized to the benefits of such concepts as world-class management techniques, benchmarking and continuous improvement, among others. Most small enterprises seem to be satisfied with their present situation. It should be categorically stated that knowledge of regional competitors regarding their management techniques, the technology used and marketing strategy, provide no good reason for complacency, but a knowledge of the global landscape is required in these areas of development. Moreover, it is important to reiterate that size is not a deterrent to achieving world-class manufacturing standards.

The foregoing issues should be perceived as guideposts and not necessarily a panacea to all the challenges faced by firms in small states. At the level of management, it may require an understanding of the global and local value chain, the vagaries, dynamism and state of flux that so aptly characterize the global market. A rehash of the corporate strategies may also require additional training and resources to lobby external forces that may likely impact specific aspects of a firm's operations. Before organizations expend resources on investment activities, they should be assured that the net returns on those incremental investments are financially rewarding in the long term.

This would prevent organizations from mistakenly reverting to fortuitous and unplanned undertakings. It would be advisable and prudent for them to seek professional assistance and guidance in conducting a feasibility assessment that could inform them about the most discrete and judicious decisions making scenarios that will inform the investment.

Chapter 12

The Paradigm of Competitiveness and Industry

The concept of competitiveness has become a widespread topical issue for many decades. There has hardly been any definitive clarity on the subject because many people have discussed it from different points of view. Generally, the discussion seems to imply that a country's competitiveness relates to its ability to maximise its comparative advantage for economic gain in engagement in international trade. The status of competitiveness is of utmost importance to small states because they lack the resources to afford the standard of living associated with developed countries. The onus is therefore on these small states to optimize the benefits that can be derived from international trade. To be able to do so, they should develop the necessary systems and tools to compete successfully with those that are more endowed with resources.

At the level of the firm, significant changes have been taking place in the way industries reorganize their production processes in response to changing market conditions. Evidently, there is a new dynamism in product demand, driven fundamentally by quality, taste, consumer income and other market dynamics. Business enterprises must therefore become agile in their response to market pressures while responding to demands for a range of goods and services as well as fulfilling the corporate goal of maximizing the profit margin.

In the context of the current global scenario and considering their limited resources and technological capability, small enterprises cannot afford to lag behind and then play catch up. Their success as international competitors will depend to a large extent on the effective transfer of technology to keep pace with modern trends in production. This will enable firms to attain a comparable level of efficiency, standards and quality and low-cost structures for production inputs, which would provide the necessary capability to foster competitiveness.

Local and regional competitive standards should not be the optimum benchmark of achievement for small and medium-sized industries. The benefits to be gained through free trade should be the impetus to pursue world-class standards as the ultimate benchmark. It should be noted that to attain a level of competitiveness comparable to world-class status, firms will require a substantial injection of resources in critical areas, as well as the restructuring of the management and production and operations systems. Notwithstanding the mediocre state of manufacturing in many small developing states, there are, however, exceptions to the generally low level. It is therefore important to conduct research that will reveal the true competitive position of these industries to determine what gaps exist regarding their current status.

There are many schools of thought as to what constitutes a globally competitive enterprise. Although not generally prescriptive, achieving international levels of competitiveness will require technological interventions driven by innovation that will integrate and invigorate processes and create new opportunities for change in keeping with global requirements in several functional areas. In this regard, the adaption of tools such as Advanced Manufacturing Technologies (AMT) and techniques that are aimed at reducing cost, improving product quality and reducing lead time will eventually forge the manufacturing process forward and create the framework for the challenges of competition.

In the past, competitiveness had been linked to product development, quality and the ability of enterprises to keep customers interested in buying their products. In today's business environment, it is a highly integrated process that must begin at the planning stage of product development and manufacturing. The concept of competitiveness is enveloped in a combination of world-class manufacturing techniques and approaches that apply to specific lines of manufacturing. Some of these techniques, though not necessarily novel, have been modernised over time in response to new technological advancements.

Privatization Option for Public Sector Enterprises

There have been ongoing debates on whether the state should get involved in private sector type commercial activities and if so, to what extent. State ownership of enterprises has its genesis in the 1960s during colonial rule by some European countries when foreign administration regulated most of the economic activities in the colonies they controlled. State domination of enterprises continued during the post-independence years for most of these small colonial states. The control of state enterprises was also driven by ideological and practical philosophies. For example, socialism, which was a direct opposition to capitalism promoted by colonial powers was a means of controlling state assets and thereby preserving economic independence in the perceived threat of neo-colonialism.

During the early 1990s, Margaret Thatcher was the leading proponent of "small" government which led to the privatization of several essential services in Britain. Today, the control of the state machinery and resources through the expansion of state-owned enterprises continues to exist in some developing countries. This policy is in stark contrast to that of reducing the size of government and instead, empowering the private sector as a major player in national development.

Privatization is an economic and political strategy designed to reduce the role of the state in public sector enterprises and increase that of the private sector in the ownership, control and management of the state's productive resources. In many cases, the choice is between allowing these enterprises to continue to operate inefficiently and thereby make losses or enabling them to become competitive and profitable. Privatization should be centrally driven by the desire to improve the efficient use of resources by shifting the productive sector of the economy into private hands.

Traditionally, state-owned enterprises have been generally known to be inefficient and uncompetitive. Even within parastatal organizations, that is, those enterprises fully or partially owned by the state and the private sector, there is a high level of inefficiency, especially where for the most part, government policies apply to a wide range of business operations including the government's commitment to state ownership and its control of the means of production. There is enough evidence to show that state-owned enterprises have become a burden on governments in developing countries around the world. They have racked up staggering subsidy costs as well as enormous debt which when added to the national debt is rather onerous. Privatizing these enterprises could relieve governments of their enormous debt burden. In many Caribbean countries, for instance, a wide range of state-owned companies including banks, telecommunication facilities and industries have been privatized over the last couple of decades and the outcome of such policy has proven to be quite rewarding for investors.

Privatization brings a host of new economic opportunities including training in new skills. Industries that were formerly run by the state and are now in private hands require greater managerial capability and a trained workforce to increase the level of competitiveness. In the post-independence period, state-controlled philosophy has now given way to new strategic thinking with regard to finding solutions to existing economic challenges including reducing the debt burden on developing states thereby allowing for more fiscal space. This does not in any way diminish the role of the state in providing goods and services to meet the growing needs of the public sector.

The neo-classical hypothesis that private enterprises bring greater efficiency and rapid growth is well advanced and accepted. The failure of public enterprises can be attributed to several factors including the pursuit of goods at prices below cost, hiring extra workers to satisfy national employment and political objectives; over-concentration of decisions that limit the flexibility of managers in everyday operations and the bureaucratization of management which provides little incentive and scope for managers to be creative and to take responsibility for their decisions. The case for privatization can be made on the following grounds:

(i) the sale of public corporations to employees, the public or other corporations.
(ii) the conversion of state-owned enterprises to private sector-type organizations with a commercial mandate.
(iii) contracting out services previously provided by the state.
(iv) the use of private sector management in state-controlled enterprise

Privatization undoubtedly has its drawbacks and may not solve all the problems which it was designed to address. What is important in the privatization process is whether it serves the long-term interests of the nation by promoting sustainability leading to economic growth and social progress.

In more recent times, many state enterprises are becoming less popular. To achieve greater levels of diversification and commercialisation, many of these institutions must be fully privatized. When an enterprise is privatized, it provides a greater opportunity for survival and growth through investment and increased competitiveness. This decision will enable shareholders to commit resources to its activities, thereby strengthening its capacity to function efficiently. The retention of state ownership will allow an enterprise to be managed within the confines of an inflexible bureaucratic structure with a piecemeal approach to the allocation of resources. Privatization engenders critical decisions concerning management, the application of new technology, and the drive towards initiatives leading to commercial innovation. This will be more effectively addressed.

Regional Industries & International Competitiveness

There is consensus that issues raised earlier including education and training, innovation, infrastructure, science and technology, business sophistication and institutional strengthening among others, are indeed pillars on which a competitive infrastructure can be established in developing countries. The World Economic Forum's Global Competitiveness statistical analysis has identified these issues as critical pillars of competitiveness. It had been long recognised that market size, business sophistication and innovation are the fundamental pillars that drive new product development and are also critical in achieving structural transformation, diversification and economic growth in small states as well as attaining levels of international competitiveness. It is, however, difficult to attain world-class status considering the size and the population of some small states, the absence of targeted R&D programmes and the financial constraints facing them.

Several indicators suggest that the Caribbean, for example, has seen a reduction in its competitiveness over the last decade. The shares of some industries in world markets have fallen, trade as a share of GDP has also fallen and the current account has deteriorated. The growth in merchandise as well as service exports has also fallen, including growth in the tourism sector, which has also declined as a percentage of GDP from 22 percent in 1990 to 18 percent in 2001. Real wages have risen in excess of productivity for many of the countries in question. Some segments of domestic production, including agro-products, are not competitive and depend heavily on high tariffs for survival. Moreover, the deterioration in the fiscal situation in some states means that macro parameters such as real exchange rates and interest rates are likely to be more volatile in the future, which is usually not good for exports.

Most small industries in small developing states are not competitive in labour-intensive manufacturing or agriculture-related industries because of the relatively high wages and the unit cost of production. Consequently, they will find it difficult to compete in a more open market regime. The services sector which is emerging as a major economic driver in small economies has the potential to be highly competitive. The examples of Ireland, Singapore, and Cyprus present dramatic illustrations of the possibilities for small enterprises in that, though relatively small in size, they possess industries that are generally world-class and are therefore competing globally.

While some diversification has already taken place in some developing countries, focusing on niche activities in some critical sectors can complement traditional activities in other sectors. This process needs to be accelerated to reverse the decline in growth that has been experienced during the 1990s. By creating a successful services sector and niche manufacturing exports, many island states can leverage their geographical location through their proximity to larger markets for export trade. In the case of the small island states of the Caribbean, their predominantly English-speaking populations and their proximity to the USA and Canada could be an advantage to trade. Their alluring natural beauty, rich cultural heritage and the abundance of indigenous tropical produce that are the envy of many countries, could form the basis for a variety of niche products that can boost export trade. There are numerous examples of service exporters that have successfully exploited these possibilities including Sandals resorts, health and wellness centres, and the burgeoning offshore education industry.

One of the aims of competition is to enable a firm to attain a level of competitiveness in performance that will match and surpass those that are considered the best in the business. What is important in manufacturing is an understanding of the organization and its management modus operandi. Each firm should seek further knowledge about the demands of an internationally competitive market, what it requires to become world-class, who are the world-class competitors within the industry and which firms are the best competitors in a specific line of business. A firm should also determine what aspects of its activities demand attention, how it is going to effect change, and what will enable it to become efficient and competitive.

International competitiveness can be considered a measure of the relative cost of goods/services from a given country. Countries that can produce the same quality of goods at a lower cost are said to enjoy a competitive edge. International competitiveness encompasses qualitative factors that do not lend themselves readily to quantification. Factors such as technological innovation, degree of product specialization, product quality, and the value of after-sales services, may influence a country's trade performance. Measures of competitiveness should satisfy three basic criteria. Firstly, they should cover all the sectors exposed to competition i.e., only those goods traded or tradable that are subject to competition should be considered. Secondly, they should encompass all the markets open to competition, and thirdly, they should be constructed from data that are fully comparable internationally.

In principle, to obtain a comprehensive picture of the competition between exporters and producers in any given markets, detailed studies should be conducted covering all categories of tradeable goods. Such studies are normally confined to aggregate measures of manufacturing output. In practice, none of the indicators available fulfills these criteria. In principle, it would be important to determine what tradable products in a country or a group of countries are internationally competitive. To do so, several quantitative variables of other competitors, production and price variability should be established. Other common competitive measures for manufacturing strategy that are not considered include short delivery cycles, dependable delivery, superior product quality and reliability, flexibility in volume changes and the ability to produce new products quickly and at low cost. This will then lead to a comparative analysis of the market performance of any given product.

Several methods have been used to determine competitiveness including productivity. The problem with productivity is that, because of the unique cultural and other peculiar characteristics including export goods and services, (including some tourism, heritage and festival products, certain financial services, music, handicrafts, agri-processed products, herbs and spices, fruits, flowers and other commodities) may be competitive in international markets even if labour productivity as conventionally defined is lagging. Real unit labour cost may not always be an accurate representation of competitiveness in that, it excludes other inputs, whereas capital and technology inputs in capital-intensive industries like tourism (hotel erection phase), mining and construction, may be of considerable importance.

Trade-revealed measures (TVM) have also been used to measure international competitiveness at the industry/sector level. One of these measures determines competitiveness according to changes in the share of world exports of the country in global markets with expanding or declining demand. Another determines a country's competitiveness by comparing a commodity's share in its exports to the commodity's share in world exports, referred to as its revealed comparative advantage (RCA). If the RCA is >1 it is taken as evidence of international competitiveness.

The concept of comparative advantage is commonly employed

to evaluate patterns of trade and export specialization. According to Heckscher-Olin theory, a country's comparative advantage is determined by its relative factor scarcity, that is, its factor endowment ratios relative to the rest of the world or a set of countries.

According to Béla Balassa (1989) since relative prices under autarky are not observable, measuring comparative advantage poses particular challenges. Comparative advantage is "revealed" by observed trade patterns. Specifically, exports could be used to reveal the comparative advantage of a particular country in the absence of data on factor costs, as the pattern of commodity exports reflects relative costs as well as differences in non-price factors that can be expected to determine the structure of exports. Béla Balassa (1965) developed the most widely used approach to analysing revealed comparative advantage commonly known as the Balassa Index. This index is essentially an index of revealed export advantage (RXA) which can be expressed as:

$$RCA_{ij} = \frac{X_{ij} / \sum_{j} X_{ij}}{\sum_{n} X_{nj} / \sum_{n} \sum_{j} X_{nj}}$$

(1) where X_{ij} is country i's export value of commodity j; $\sum_{j=1}^{J} X_{ij}$ is the total export value of country i; $\sum_{n=1}^{N} X_{nj}$ is the export value of commodity j for a set of n reference countries; $\sum_{n} \sum_{j} X_{nj}$ is the export value of all j commodities for a set of n reference countries; and RCA_{ij} is the revealed comparative advantage of country i in commodity j. RCA_{ij} greater than 1 indicates that country i has a comparative advantage in the export of commodity j in the market of a set of reference countries; while RCA_{ij} less than 1 indicates the country possesses a comparative disadvantage.

This index can be used as a basis for providing information concerning a country's competitive position relative to a group of countries within a grouping such as those in the CARICOM Community or the Pacific Islands Forum. Such information could identify commodities in which the countries in question are internationally competitive. Countries are deemed competitive in a product if they hold a comparative advantage in at least three of the five years being investigated.

Many of these SIDS are competitive in a wide range of broad product categories. Almost all of them hold comparative advantages in beverages, while a significant number are also competitive in crude fertilizer, petroleum and petroleum products, and essential oils and perfumes. For example, beverages are comprised at the 5-digit level of waters, including natural or artificial mineral waters and aerated waters; waters containing added sugar; vermouth and other wines of fresh grapes flavoured with plants or aromatics; sparkling wine; wine of fresh grapes (other than sparkling wine); whisky, spirits obtained by distilling grape wine or grape marc; rum and tafia; gin and geneva, spirits, and distilled alcoholic beverages.

There are obvious shortcomings to the aforementioned measures in that, they can be affected by a variety of factors that have little or nothing to do with a country's intrinsic competitiveness in the industry or sector concerned, such as available supplies, transport, marketing, niche products and destinations and product diversification on world markets. TVMs have been applied to selected countries with comparisons made to some other countries. These measures have not always given consistent results but suggest that most export commodity groups are not internationally competitive, while a limited number are 'safe' and have expansion potential. Nevertheless, it is possible to give some depiction of the situation regarding competitiveness at the sector/industry level.

Regarding Caribbean economies, for example, traditional industries such as sugar, rum and mining (bauxite and alumina) industries hardly had any competitive advantage. To some extent, the bauxite industry as in the case of Jamaica thrived for some time because the metal ore lies close to the surface which gives rise to its easy extraction, while in Guyana, the competitive advantage lies in the fact that Guyana was the world's leading supplier of so-called calcined bauxite, a high grade of the mineral used for lining steel furnaces and other high-temperature applications. These advantages were lost because of several exogenous as well as economic factors. It is important to note that because of the importation of heavy equipment and additives, these industries have become import intensive. Moreover, there have been concerns expressed over the by-products from the discharge from some of these industries.

Chemicals, cement, furniture and metal products are important subsectors that emerged as a result of the import substitution policy that never gave countries that adopted it a competitive advantage. The agro and agri processing sectors have been historically linked to the agricultural sector and like other light manufacturing activities have faced the same constraints discussed earlier. From the supply side, one of the perceived advantages of these industries is the availability of raw materials.

Since the Caribbean, Asian and African states are such a small proportion of the world market, they essentially faced an infinitely elastic demand for their exports. In other words, they could sell all that they could produce at the going market price. As a result, these small states could specialize in areas of manufacturing that exploit their most abundant resource (which is the people) and export their higher value-added output to more advanced economies. On the contrary, the quality of the raw material input, and the quantity demanded can hardly be sustained because the supply chain is disorganised and underdeveloped. Many of these manufacturing industries have been operating inefficiently and have not integrated important subsectors of the national economy.

There are important lessons to be learned regarding how some small enterprises have developed highly successful world-class businesses. One such story comes from the small Chinese town of Qiaotou. Hitherto its global button fame, nothing was known of this insignificant town. The story of Qiaotou began in 1980 when three brothers from the town started a business by picking buttons off the street. Twenty-five years later, Qiaotou has transformed itself from a farming village into a manufacturing powerhouse producing over 60% of the world's buttons and 80% of the world's zippers.

The potential for explosive growth is distinctive to manufacturing. As manufacturing activity expands, instead of running up against shortages of land or resources that inevitably constrain the growth of agriculture or the extractive industries, there can be enormous benefits from economies of scale as the unit cost of production falls. Prior to globalization, although such cost reduction helped manufacturing to expand, the size of the domestic market constituted a constraining factor especially in small low-income countries where the tiny market for manufacturing limits the scope for benefitting from economies of scale. Now that markets are global, this constraint no longer exists. If a country can find a niche in the global market it can scale up almost any commodity without limit, as demonstrated by Qiaotou.

A study on "Competitive Priorities of Manufacturing Firms in the Caribbean" was conducted by the Centre for Production Systems, University of Trinidad and Tobago, and Sobey School of Business, Saint Mary's University, Halifax, Nova Scotia, Canada. There were several interesting findings from the study, which underscored the competitive nature of Caribbean firms that need to be highlighted.

The five competitive priorities of the study included cost, quality, flexibility, delivery, and innovation. The results revealed that among the five competitive priorities studied, the issue of cost was most strongly emphasised by the respondents. This is contrary to modern-day thinking that competition based on lower costs is no longer effective in today's marketplace. The emphasis on cost is perhaps linked to the prevailing economic environment in the Caribbean. In the 1990s, the era of preferential access to European markets for Caribbean products was ending. In the aftermath of preferential market access, there has been an influx of imported goods from low-cost producers from other countries.

Some manufacturers have been unable to produce goods cheaper than the price of equivalent imports from extra-regional markets largely because of constraints in material and labour inputs as well as the high costs of energy and transportation. As a result of these constraints, manufacturers in the region were forced to look toward low-cost manufacturing. The study indicated that manufacturers are focusing on improved labour productivity and operating efficiency as means of reducing cost, as opposed to sourcing low-cost materials, and reducing overhead costs.

The emphasis was therefore on controlling internal factors such as labour and equipment performance rather than external factors such as material cost. In a region where some countries have experienced modest growth and where others have declined, local consumers are very price-sensitive and as such cost appears to be an order winner for Caribbean manufacturers. This is in contrast to more developed economies where it is an order qualifier.

Quality is ranked second as the most important competitive priority. The result shows that 3 of the top 5 priority items relate to quality. It is encouraging to note that Caribbean manufacturers are placing more emphasis on developing capacity in quality. This is a most important factor in competitiveness, particularly for small producers with low production volumes, which is characteristic of most of the firms in the study. Products that conform to quality standards and are customer focused have the potential to lead to brand loyalty, particularly so for a region where local consumers still perceive imported manufactured goods, mainly from Europe and the USA are of higher quality than locally produced goods. As such, firms that emphasize quality are more likely to build a reputation by changing this perception and therefore gain market share and sales growth.

Concerning the age of the firm, mature firms strongly emphasize cost and innovation. Older firms appear to have realized over time that to be competitive they need to focus on cost reduction and greater labour productivity. With experience, these firms have also recognized the importance of product innovation and the timely introduction of new products to the market. Since most of the firms in the study cater to the domestic market, which is fairly small, companies that introduce new products or make incremental improvements to existing products are more likely to win orders. It is not entirely surprising then that the results show that firms with high annual sales place more emphasis on the competitive priorities of cost, flexibility and innovation when compared to firms with low sales.

The survival of manufacturing in the Caribbean and other small states will depend on the extent to which manufacturers can remain competitive despite challenges such as high labour costs, high energy costs, high financing costs, inadequate infrastructure and insufficient government support. To be locally and globally competitive, manufacturers need to maximise their competitive advantages by exploiting their manufacturing capabilities, notwithstanding the challenges they face because of their relatively small size.

Approximately 50% of the manufacturers in the study implemented quality assurance programmes while only 18% have obtained ISO 9000 certification. Caribbean manufacturers, therefore, need to focus on improving practices in quality management systems such as statistical process control and total quality management (TQM) as methods to achieve consistent quality. In addition, because of the high operational costs, cost reduction and efficiency systems such as Six Sigma and lean manufacturing should be considered as options to support low-cost manufacturing.

Another strategy recommendation is that Caribbean manufacturers should focus on niche manufacturing. Whereas large economies like China and Brazil can compete on low prices and large-scale production, economies of most small states may have to select competitive niches that will allow them to sell small volumes at high margins. Most small states in the Caribbean, for example, can only compete profitably in the manufacturing sector by developing and exporting to niche markets where consumers are not price-sensitive and products are not based on low labour costs. An example of this is pharmaceutical manufacturing in Puerto Rico, which was once the fifth largest in the world for pharmaceutical manufacturing. Niche product development, marketing and strategic thinking have been key features behind the success of SM Jaleel, both in terms of the products they developed and how and where they are marketed.

Currently, the emphasis of small enterprises is not based on capabilities in flexibility and delivery, because it requires greater use of automated and flexible production processes to promote flexibility and delivery. It is important to note that lack of access to capital makes it difficult for firms in small developing countries to invest in technology to achieve flexibility in their operations. Governments could therefore be encouraged to provide financial incentives such as low-interest loans to manufacturers to invest in automated manufacturing systems.

In a global market that is highly dynamic and competitive, the issues of efficiency and competitiveness in enterprises should be addressed with urgency. Recommendations toward competitiveness should be perceived as guideposts and not as dictates. Implementation of any set of recommendations requires dialogue at all levels of the management spectrum. It may necessitate a rehash of the corporate strategies which may require additional training and resources to lobby external forces that impact specific aspects of the firm's operations. Before organizations expend resources on any set of activities, management should be assured that the net returns on investment are financially rewarding in the long term. Organizations do not have to revert to fortuitous and unplanned undertakings.

It would be prudent and advisable for enterprises to seek professional assistance and guidance in conducting exercises that could improve their status as well as to inform them about the most discrete and judicious decisions to foster confidence in the undertakings of the organization.

High-End Value Added

The term "value-added" can be described as any enhancement a company makes to its product or service before offering it to customers. An enhancement can be in the form of an extra special feature added to the product or service to increase its value and thereby attract customers to purchase the product. For example, a firm manufacturing a product that is considered homogeneous is differentiated from those of its competitors in a few ways including an additional feature or add-on that gives it a greater perception of value to its customers. A company may add a brand name to a generic product, or it may produce a product in a way that no company has thought of before. Arithmetically, the value which is added to a product is the difference between the price of a product or service and the cost of producing it.

Manufacturing is the transformation of materials into new products. Such transformation can take place through a physical, mechanical or chemical process using machines or hands respectively. The point was established earlier that small states including the African, Caribbean and Pacific (ACP) states traded raw materials with their former colonial masters in Europe in their primary form. In so doing, they have not given any consideration to developing industries that will allow them to add value to these products so that they can attract higher prices on the market.

The manufacturing sector in many small states has suffered due to high production costs and small production volumes. In the Caribbean for example, high energy and wage costs coupled with less competitive financing costs predispose regional manufacturing firms to be less competitive than those in other middle income developing countries. Wage levels have also increased faster than average labour productivity in the region. Value-added in regional manufacturing is usually low, as production often entails the assembly of foreign intermediary inputs with little local content and industry linkages. These factors have generally slowed the growth of manufacturing output.

Many critical domestic factors affect the competitiveness of producers in small states. These include low productivity especially labour productivity; small market size; low investment in research and development for product innovation; a relatively underdeveloped institutional framework to support business development and sometimes a convoluted business environment. The increase in competitiveness is also predicated on knowledge and information rather than on traditional factor accumulation.

The quality of research undertaken and the education and training systems in some jurisdictions are often unrelated or inappropriately structured to the needs of the industrial sector. Hence, these systems need to be revamped to address the demands of the new economic realities. The increasing plethora of international industrial and business standards now driving companies' ability to participate and compete in markets means that business enterprises should modernize and restructure their processes and management systems to consolidate and improve their growth prospects. The ability to raise financing on competitive terms has also been a major problem for some firms, especially small firms in non-traditional sectors such as the professional services sector and the entertainment industry.

It must be recognised that the transition to niche manufacturing and high-end services could be an onerous undertaking for small island economies without the necessary skills and resources to do so. Other small island developing states have managed the transition successfully from which valuable lessons can be learned. Ireland had a dramatic transformation from a largely low-cost agricultural and manufacturing economy to a service, pharmaceutical and information technology economy that has achieved a three-fold per capita GDP increase since 1970, accelerating from negative growth to the fastest in Europe.

Through coordination between the private and public sectors, Ireland targeted and attracted companies in key sectors. In turn, the government created the necessary infrastructure and investments in education to supply the skillset required in the engineering and computer science sectors. Ireland went from having the worst schooling record in the European Union to graduating proportionally the highest number of scientists and engineers in the European Union (EU).

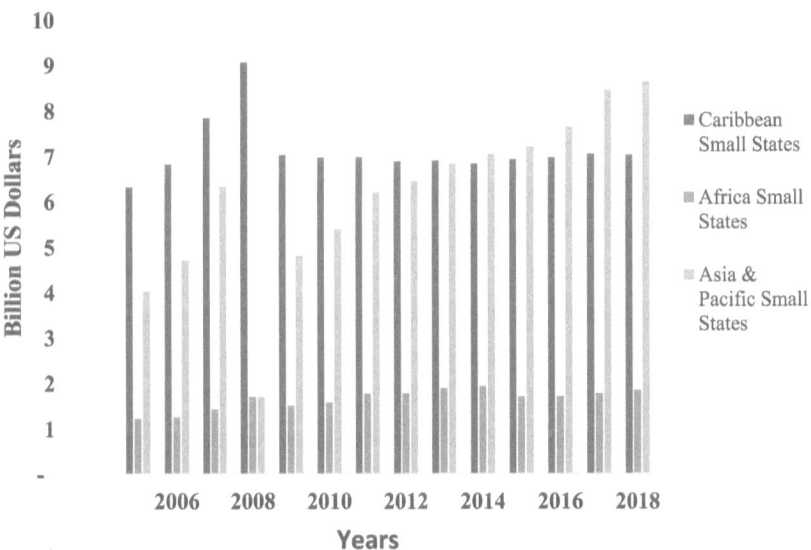

Figure 12.1: Value Added Manufacturing in Small States

Niche Manufacturing in Small Island Economies

In the context of business, a "niche" is defined as a specialised but profitable corner of the market, where specialization alludes to a measure of product differentiation. Hence, a niche product derives its value from differentiation. Niche products target a relatively small segment of the market. Similarly, a niche market is considered one that consists of relatively few customers willing to pay premium prices for goods and services that meet their specialised needs. From this perspective, niche manufacturing may be considered as the manufacturing of a product, which is oriented towards serving a final generic purpose but is marked by an ultimate form of differentiation. Niche manufactured products may be differentiated in terms of their input source, method of production, brand name, country of origin, price and inherent characteristics (for example, durability or value-added functions).

Market size may not necessarily be considered a characteristic of niche manufacturing, as niche products may command a large market share. In addition to the size of potential demand, such products tend to be inelastic and highly priced (price-insensitive) and driven by differentiation. Niche manufacturing also has the potential to boost foreign exchange and enhance the productive base of a country. Given its unique characteristics, it can be an option for diversification of a given sector such as the agricultural and manufacturing sectors, which may need to respond readily to the dynamic needs of the market.

The requirements for niche manufacturing can be less capital intensive. Smaller production plants that will provide the flexibility to adapt quickly to production patterns and the ability to incorporate unique features into products that encapsulate distinct features of the country are advantages of niche manufacturing. Most exporters tend to be relatively large firms operating as limited liability enterprises. Most niche manufacturers tend to be small unincorporated businesses. One of the first goals in developing a successful niche manufacturing enterprise in a specific region is the reorganization of these enterprises into limited liability entities. This would allow those businesses to benefit from the increased human and financial resources.

Most small developing states continue to focus on traditional industries to drive their economies targeting traditional markets. In many instances, agriculture is still the main economic driver followed by tourism. Global markets have become dynamic with a variety of new products that act as a substitute for traditional ones. For example, there are sweeteners on the market that can compete with the sugarcane grown in many tropical countries.

To explore new market opportunities for existing products, a combination of strategies may be useful to revive the fortunes of declining enterprises. This is especially important if enterprises want to avoid direct competition in markets where larger and more established competitors dominate. For example, niche marketing techniques could be useful in reviving the fortunes of ailing traditional industries. It will also allow an organization to concentrate on a specific market need, which is often ignored by larger competitors.

Symbiotic relationships with firms of similar interests can be developed to facilitate the mutual sharing of information, technology, market intelligence and training. Small firms within a given proximity of leading firms can benefit from economies of scale through wider cooperation. These firms should examine the possibility of jointly establishing a "cluster of firms" to support shared infrastructure (management services, equipment and technology) that the individual firm could find uneconomical to finance.

Niche Manufacturing; Challenges and Opportunities

There is no doubt that niche manufacturing can enhance the international competitiveness of small enterprises. Several studies have pointed to the fact that the SME sector, which is a component of the manufacturing sector, has contributed significantly to GDP in small developing states. A thorough analysis of the manufacturing sector emanating from numerous studies suggests that there is tremendous scope for the development of niche products and improvement in existing ones to enhance the marketability and export potential of the product offerings within the sector. It has become necessary to create backward and forward linkages with key sectors like agriculture and tourism.

Private sector investment is at the heart of economic growth and poverty reduction, creating opportunities and employment, increasing the choice of goods and services as well as lowering their cost and providing a source for tax revenue. Generally, economic growth is higher in countries that have higher investment rates. Growth has been shown to have a long-term effect on wealth creation, employment, and poverty reduction. Generating investment requires a good enabling environment that fosters productive private sector investment by removing unjustified costs, risks and barriers to competition. A good investment climate unleashes competitive forces that stimulate innovation and productivity.

Regional integration could be a useful approach if small states can take advantage of labour mobility, more open trade, joint investment promotion and the effective functioning of institutions, regulatory cooperation, and the like. The benefits of regional integration could be enhanced through the reduction of external tariffs, which would reduce trade diversion.

There is a range of untapped possibilities for industrial development and in particular, niche manufacturing in many small developing states. By stimulating innovation, the application of simple technologies can promote self-sustaining development opportunities. Development in process technology, for instance, can result in new market opportunities in material science through the production of new materials for industrial application. Technological breakthroughs made by research institutions could reasonably assume commercial significance only when they result in offering a product to the market. The viability of such commercial ventures depends to a large extent on the financial support given by the private sector and agricultural institutions supported by governments.

One of the lessons entrepreneurs have not fully learned from the industrialised countries is how to establish successful enterprises from indigenous commodities. For example, New Zealand has internationalised the Kiwi fruit through proper marketing. The cosmetic industry uses the derivatives from the Aloe Vera plant in various mixtures. It must be recognised that Aloe Vera is grown abundantly throughout the Caribbean and not much of this product is used for commercial purposes. For many years, the general use of cannabis was prohibited in most countries. Some Asian and Middle Eastern countries have laid down severe penalties for the use of illicit drugs including imprisonment for several years. In recent years, countries including Canada, Georgia, South Africa, Uruguay and many states in the USA have adopted policies to decriminalize the use of the drug for recreational purposes.

In the last couple of years, the ongoing discussion and debate in most Caribbean countries about the use of Cannabis (Marijuana) for medicinal purposes has been gaining momentum. Recent developments emanating from documented cases about the success of its use to treat isolated medical cases have triggered widespread discussion and interest in the drug. Some developed countries have enacted legislation to decriminalize the use of marijuana primarily for medicinal purposes, albeit under a legally controlled environment. It is estimated that the drug has a market potential of hundreds of millions of dollars in the long term. Research is currently ongoing regarding the range of other uses to which cannabis can be ascribed.

Through the expansion of the agro and agri processing sub-sector, there are opportunities to diversify the ailing Agriculture Industry in some developing states through several value-adding activities driven by innovation. Presently, the global focus on health and wellness has created a derived demand for products such as fruit nectars and beverages from fruits and vegetables that are organically grown with minimal artificial additives. Apiculture, alternative medicine, as well as specialised products from herbs and spices can form new areas of development for specialised products from the Caribbean and other regions' rich indigenous tropical resource endowment. Enterprises in SIDS should maximise their indigenous resources as the base for their niche product development. It was stated earlier that there is a large variety of unutilized tropical products that could form primary raw materials for many new marketable products.

Within the niche manufacturing sub-sector, there are opportunities and scope to develop authentic ethnic foods for overseas markets by providing value-added food products of higher quality than their competitors. The Caribbean region for example, is replete with a variety of indigenous raw materials with possibilities for new products with commercial applications which can support the tourism, agricultural and art and craft sub-sectors. Many opportunities could emerge for the development of niche markets for exotic fruits, flowers and oils for the cosmetic and fragrance manufacturing industry.

Other possibilities that require investigation could include the extraction of natural fibres for a wide range of commercial applications from the pseudo stem of the banana as well as derivatives from plants such as the Aloe Vera. Sea Island Cotton also has applications as a high-quality fabric for specialised use in the garment and craft industries. A variety of other natural products holds tremendous potential for the development of new products that can strengthen the niche market and thereby become niche market leaders. In recent years, natural fibres have been the subject of intense research and development. Apart from their tensile strength and durability, they are known to have tremendous environmental benefits. The fibre from the Manila Hemp also called Abaca is grown mainly on small multi-crop farms in the Philippines, where there are some 200 varieties in cultivation.

This plant could produce fibres similar to jute, flax, kenaf and hemp. The Ababa plant is reputed for high-quality fibre, which is resistant to saltwater damage, has a long fibre length of up to about 3 metres and has excellent tensile characteristics. The Abaca fibre is pulped and processed into speciality papers including high-quality writing paper and currency notes. It is reported that about 30% of the Japanese yen banknotes contain Abaca. When stripped from the pseudo stem, Ababa is made into a range of products including ropes, cordage, high-quality resistance paper, yarns, abrasive backing paper, tea bags, attractively woven fabric, handbags, purses and shoes.

Similarly, the pseudo-stem of the banana plant, which is a species of the Abaca plant with similar properties can be the subject of further research with the aim of future commercial uses. Exports of Abaca products are an important earner of foreign exchange, with two-thirds of export going to the US, Singapore, Australia, Malaysia, Thailand and Europe are also important markets.

Although exports of bananas to European markets from ACP countries (that are predominantly small developing states) have declined significantly. In Guadeloupe, waste from the banana (plant and fruit) is transformed into biogas which is a non-polluting renewable energy source as well as feedstock to produce biofuel. These by-products can be used for heating, cooking, automotive fuel, and electricity generation. The valorization of plant wastes can also be an important sustainable alternative for small states in the field of energy, fuels and other environmentally sustainable products.

Some small island states have been using indigenous raw materials to establish niche products that have been enjoying success in the market. It is instructive to note that entrepreneurs in two of the small island states of the Caribbean viz. Grenada and St Vincent and the Grenadines have been manufacturing a dark organic chocolate called the Grenada Chocolate and Vincentian Dark Chocolate respectively. The cocoa plant is a tropical indigenous plant grown locally on both islands. The cocoa plant is grown naturally without the use of any agro chemicals. The chocolate product is manufactured from cocoa beans into a delicate product that is used for making chocolate. Traditionally, the beans have been exported or grounded and rolled into cocoa sticks and used locally as a hot chocolate drink.

The rapid depletion of the world's resources, notably the extinction of endangered species of aquatic life by the indiscriminate action of man, exhausting supply of tillable soil and the increasing world population growth which has led to famine in some developing countries should be seriously addressed. The application of appropriate technology in these areas of interest can provide the solution to sustainable livelihoods and offer an alternative option to feed the seven billion people on planet earth.

Small developing states cannot be seen as merely surviving in the face of serious global challenges. Mankind cannot afford to self-destruct by using antiquated technology to increase food production and enhance sustainability. Many of the foregoing value-added products are not necessarily novel but can form part of the technological niche transformation of the economies of small developing states that are designed to provide alternative livelihoods for disadvantaged people.

Small island states generally lack the requisite natural resources and capital for investment. They have very small market sizes and are isolated from their major markets, which create logistical challenges. Local entrepreneurs, therefore, tend to focus on low-risk business activities such as wholesale and retail trade or services. Achieving economies of scale is one of the main drawbacks associated with these business activities. These areas of activity can act as a drain on the country's foreign exchange reserves and provide limited opportunities for labour which can leave the islands highly vulnerable to external shocks.

The private sector in small states remains a key driver of economic growth and wealth creation, employment, investment, and a source of finance in building capacity and human capital. For this reason, the private sector must continue to play a leading role in industrial and national development. It must, however, continue to provide proactive leadership in the promotion of sustainability in several development areas such as manufacturing, energy and mining.

The private sector must continue to create opportunities from its slender resource base that will augment its value-added ambitions, and at the same time provide risk mitigation measures for unforeseen outcomes in several critical developmental areas. These goals cannot be achieved unilaterally, but the private sector given its pool of resources should work collaboratively with the government, civil society or both, to provide solutions to many cross-cutting existential issues. Notwithstanding the role of the private sector, it is the society/community which constitutes the market that determines the success of an enterprise. For this reason, the private sector has a social responsibility to contribute to society.

Over the years, evidence has shown that several micro and small businesses have been competing against one another by duplicating the same product line using the same strategies and targeting the same markets. Most of these entities primarily located in rural districts grew out of a need to address poverty, which is often prevalent among the rural poor. Many proprietors or entrepreneurs are untrained and lack the financial, technical and managerial skills to develop the business into sustainable and profitable entities.

For the most part, most of these low value-added products are typically regular everyday products such as plantain, cassava and potato chips, pepper sauce, jams and jellies and fruit juices sold along the roadsides and at social events in many towns and villages. Most of these micro and small producers concentrate on the local market, which consumes a limited amount of these products. In most cases, there is little or no motivation and strategy to target regional or international markets because the quantities of goods produced are often limited and insufficient to satisfy regional demands.

Historically, Produce Chemist Laboratories (PCL) were established in the early 1970s in several small Caribbean islands with support from the British Government. The laboratories were initially intended to be Research and Development (R&D) units aimed at encouraging potentially small entrepreneurs to take the formulas they had developed and manufacture products on a commercial scale. The operations of these laboratories were later placed under the control of Marketing Corporations within the public sector. Over time, the initial goal of these facilities changed and redirected their activities towards making them self-sufficient. This meant that there were few research and development activities, a decision that subsequently led to the closure of most of the labs in the 1980s.

Fundamentally, some of the objectives that guided the work of the PLC were replete with developmental shortcomings. The PLC pursued a "technology push" rather than a "market pull" approach which created some difficulty for the institutions. After expending scarce resources toward developing product formulations there were no entrepreneurs who were interested in commercializing them. It is within the foregoing context that the concept of niche manufacturing was promoted. "Business as usual," is no longer an option, but sustainable business practice and competitiveness can go hand in hand. There are many incentive programmes available in some countries to support niche manufacturing although many of them have not yielded the intended results.

Chapter 13

Development Strategies and their Relevance to SIDS

In attempting to discuss the industrial development options for small developing states, it is important to do so comparatively with other relatively small island states that have homogenous characteristics from a historical, geographical, and political perspective. Such comparisons will be made within the context of the policies employed in small states that have experienced advanced economic development and could be used as a model for the development of other small states. There are many lessons to be learned from island states like Singapore, Hong Kong and Taiwan. Their economic success can be used as a model for other developing economies to emulate, albeit with caution.

The newly industrializing economies (NIEs) share several characteristics with industrial latecomers in earlier historical periods. Their development model could be successful depending on several inherent factors including the human resource, resources endowment of the recipient country, cultural attributes and policy imperatives. The political system and the policies of the government adopting the model could impact its success or failure. The essence of such a model should speak to its transferability within a comparable environment so that it can yield similar success regarding growth and equity.

Many policymakers from small developing states are generally intrigued when they read of the success of some relatively small developing countries such as the NIEs from the East Asian bloc. The success of relatively small economies as evidenced by the general standard of living, GDP per capita and the diversity of economic activities merit investigation. Although they are somewhat bereft of natural resources like many other small island developing states, their economic success is based on a diversified and complex process of development not germane to many states. Proponents of the adoption of this model of development should carefully and analytically consider the economic and sociocultural context of such development models because once a model is disaggregated, it can lose its cumulative effect and ultimately reduce the economic returns.

By definition, a "Model" is an exemplary system, device, etc. that merits replication and can serve as a benchmark for performance efficiency. A development model is different from a standard prototype model. For example, a prototype has distinct physical properties and features while a development model is influenced by domestic as well as exogenous factors. Development models are not generic and therefore cannot be adopted in their entirety by every county or state. Factors that contribute to the success of a model in one country may vary from those of another country. Consequently, the socio-cultural and economic realities of developed countries may be incompatible with those of small developing states.

Export-led growth fundamentally requires that emphasis be placed on developing a thriving industrial climate in which manufacturing (including enclave industries) should play a key role, supported by industrial policies that inform and guide the relevant fiscal policies of the government. For example, some countries are endowed with mineral wealth, favourable conditions for agricultural development and attractive marine resources. The products derived from mining, agriculture and fishing should form the nucleus of those countries' export development. Before a model can be considered, there should be massive consolidation of a country's industries and export trade.

Post-war economies of Germany and Italy have provided the lead in their model of export-led growth for countries like Chile, China, Singapore and other emerging economies. The key factors to economic success are hinged on the following:

(i) Privatization of the economy
(ii) Privatization of the social security system
(iii) Sound fiscal management
(iv) Independent monetary policy
(v) Political stability

Comparative Experiences; The Newly Industrializing Economies

The experience of the first-tier Asian NIEs which include Singapore, Taiwan, South Korea and Hong Kong commonly referred to as "Asian tigers" experience has been promulgated as a model for small developing countries to replicate. What is hardly ever said, is that the Asian model is a classic case of Sir Arthur Lewis' model of export-led industrialization. The newly industrializing economies of Asia emphasised export-led growth as the path to development, whereas many other small developing states including ACP states focused on import substitution.

Import Substitution Industrialization (ISI) policy was initially designed to replace manufactured imports with domestically produced substitutes. Several small island states implemented the policy to reduce their dependence on imported goods and simultaneously create employment opportunities. Incentive schemes were created to encourage the development of small enterprises. Incentives such as tax holidays, provision of credit schemes, the development of industrial estates and the implementation of tariffs that were designed to protect local industries.

The emphasis on export-led growth in NIEs has provided the framework for competition, foreign direct investment and adaptation of technology and learning by doing. These factors, combined with sound macroeconomic policies provided the catalyst for competitive advantage abroad. In the case of Singapore, the economy was transformed from an uncompetitive import substitution-based economy in the early 1960s to a vibrant, diversified, competitive export-led economy by the mid-1980s.

In Singapore, like the other Asian tigers, there was also a movement away from economic control and state planning to market price signals and competition forces. Modern day Singapore is viewed as a stable economy with sound macroeconomic policies established on market forces. Export-led growth fundamentally requires a thriving industrial climate in which manufacturing (including enclave industries) should play a key role supported by industrial policies that inform and guide the relevant fiscal policies of governments.

Most development models have industrial relevance and application of one sort or the other. In the case of small developing states, they must decide on their industrial priorities. It should be clear what their productive strengths are, and whether there is adequate infrastructure to support those undertakings. Such decisions should be based on the strength of a country's resource endowment from which will emanate its industrial activities. Other supporting factors include regulatory and fiscal decisions as well as the cultural preparation and adjustments that are critical and should be treated as a matter of priority.

In the NIEs, for instance, there is an enviable work ethos, which is based on discipline and commitment. This approach to work and production is largely cultural and is generally supportive of the ideals of industrial development. Culture can significantly impact development models. Traditional oriented people will resist new technologies and training that are prerequisites to a given model to protect their culture. On the other hand, more futuristic cultures will strive for new technological development, retraining and other necessary interventions in society that are designed to bring about social and economic progress.

For many decades, the newly industrializing economies (NIEs) including Hong Kong, Singapore and Taiwan have adopted an Export Oriented Industrialization Policy as their development model. This approach was based on traditional labour-intensive products and later, the emphasis shifted to more complex and sophisticated industrialization based on capital and technology-intensive manufactured products. Many of these East Asian countries prioritized the promotion and development of advanced technologies and innovation to enable them to maintain a competitive edge in the face of fierce competition from their industrialised neighbours. They have subsequently moved beyond export-oriented industrialization of traditional labour-intensive production to more complex and sophisticated export-oriented industrialization based on the export of capital and technology-intensive manufactured products.

There are obvious lessons to be learned from the NIEs experience. Generally, what is considered an economic model could be often difficult to replicate in other countries where there is a huge disparity in the distribution of wealth. Furthermore, many countries have built their economic success over many years of struggle, sacrifice and sometimes amidst territorial conflicts that have provided experiences from these unique circumstances. They have learnt difficult lessons of trust, hard work and commitment that have been ingrained in their society and become an integral part of their cultural identity.

The main element of economic success is the strategy that is being devised in adopting a model. For a model to be reproduced elsewhere, the central elements of a strategy should include (1) A set of enabling incentives, which will make it profitable to invest in activities that will utilize the abundant skilled labour. (2) Extensive government intervention in the economy to compensate for weaknesses and distortions that exist in market economies, including measures to further encourage new technologies and innovation and to offset the cost of breaking into the world market. (3) Increase scope to counter competitive pressures, primarily by providing strong incentives for industries to compete in the world market. This will foster efficiency through the use of technology supplemented by some labour-intensive activities, because in labour-abundant countries, labour intensity cannot be fully displaced. (4) High return on investment particularly in an environment in which arbitrary personalistic decisions are limited. Such a strategy will encourage the growth of labour-intensive manufactured exports.

Notwithstanding the concerns regarding the adoption of the aforementioned model, there are important lessons from which small developing states can learn in their quest for economic progress.

However, there are also several fundamental steps towards establishing a state of readiness. In this context, small enterprises will be required to adopt elements of World Class Manufacturing techniques designed to improve efficiency and productivity as well as to develop the relevant entrepreneurial capacity. The issue of work attitude can be addressed through education and training geared towards current industry needs. Indeed, the relevant regulatory and fiscal framework should be established to support these endeavours.

Lessons from Comparative Analysis

The path to industrial development success vis á vis the historical analysis of the economic growth of small island states in the Far East, though commendable, presents its own set of challenges and opportunities. Though not comparable in size, their historical and colonial status is not dissimilar to that of many of the small island developing states of the Caribbean, Africa and Asia, many of which had been colonies of some European countries for many centuries. Like Hong Kong, Singapore and Taiwan, their colonial status has always been regarded as a stranglehold on their economic development. They have, however, overcome this hurdle, firstly through independence. Political independence has to a large extent enabled these countries to shrug off the hegemony of colonialism and imperialism.

In their post-independence era, these Asian states with a colonial history got the opportunity to chart their own path to industrial and economic development. Their economic progress evolved through difficult phases of development through which they gradually and successfully navigated by employing progressive economic strategies and policies that have made them economic giants today. In the meantime, other SIDS continue to lag behind, languishing for many years in economic stagnation. What is elucidated in all these cases is the shift in industrial focus from import substitution to export-oriented industrialization.

Moreover, Hong Kong has moved beyond export-oriented industrialization of traditional labour-intensive products to more complex and sophisticated export-oriented industrialization based on the export of capital and technology-intensive manufactured products. During the eighties and nineties, many small developing states embraced import substitution industrialization as the sole industrial policy in support of their domestic industries. In the process, they have built high tariff barriers against the free flow of foreign goods and services.

This inward-looking policy perception gradually changed as the countries opened up their economies for trade in compliance with the WTO's liberalised trading regime, driven by open competition. The policy of export-oriented industrialization has never been given full effect because the countries that embraced it have never established the requisite instruments to transform the industrial sector to meet the demands of the market. Some governments have been rather slow to adjust their industrial and trade policies to take account of these new developments and paradigm shift in the global market.

Some states have advanced other reasons not to pursue the IS policy. For example, the government of Fiji thinks that the import substitution policy entails considerable costs and is not viable for small states. Hong Kong on the other hand has regarded this policy as one of protectionism and failure. Singapore considered its size as being too small for any serious agricultural development. This thinking has resulted in a shift away from agriculture as its main economic activity. It has also recognized the role of manufacturing and other forms of industrial development as key to their economic success. Some countries have not made the requisite economic adjustments and are still dependent on agriculture as the main export trade. In other words, they have embraced in some form or the other, the same policies that other small states considered to have retarded their economic progress.

Hong Kong, Taiwan and Singapore have been able to capitalize on their physical advantage with respect to their proximity to the Asian market and their ideal location (notably Singapore and Hong Kong) as a gateway to South-East Asia. In this regard, they developed powerful and influential financial centres that command global attention. It was pointed out earlier that many Caribbean states have not been able to capitalize on their geographical proximity to the North American market in the same way as these Asian countries have done.

All the Asian states in question developed strong alliances with industrial giants such as Japan and China, which have enabled them to establish industries of world-class status. Perhaps the market limitations of the Caribbean Community, for instance, as a regional grouping should serve as a stimulus for targeting markets within its geographic proximity and ultimately the global market. Whereas the Asian states have successfully penetrated other markets within that hemisphere, other SIDS have not developed strategic alliances with developed countries within their hemisphere.

The state of Mauritius is an outlier in Sub-Sahara Africa. During the decades of the 1980s and 1990s, the Mauritius economy underwent a remarkable transformation from a monocrop sugar producer to become one of the major exporters of manufactured products in sub-Saharan Africa. Its success in the export of garments can be attributed to a market-oriented development strategy.

The industrial achievements of the NICs could not be possible without an adequately trained workforce which aided the development of professional competence. The governments of the NICs prioritized the promotion and development of advanced technologies and innovation to enable them to maintain their competitiveness in the face of competition from their industrialised neighbours. They have also invested significantly in education and training in the fields of science and technology and engineering to pursue their sophisticated export-oriented industrialization. More importantly, they have been able to retain their trained professionals by creating enormous opportunities for investment in many areas of development including information telecommunication technology and the services sector. The opposite holds true for other small states where training is not rationalised through proper manpower planning and given top priority particularly so in respect to fulfilling the needs of the industrial sector.

It cannot be disputed that many small developing states lack the resources necessary to devote to meaningful investment in science and technology and engineering. It is for this reason that quite often, there is a dearth of professionals in these areas. At the macro level, there is little evidence that any serious manpower planning is done, hence, scholarships and other training opportunities are often provided or approved by the government solely on the academic choice of students and not necessarily based on a desire to address the long-term needs of the country. Consequently, professionals emerge fortuitously, thereby creating a dearth of technical and academic professional expertise in key sectors.

Against the foregoing, the most forward-thinking policies should be geared towards the following:

(i) Focus on higher-value manufacturing;
(ii) Promoting niche clusters;
(iii) Development of the physical infrastructure as well as its soft infrastructure such as ICT; technology transfer and innovation;
(iv) Capacity building.

The Asian states in question recognised the importance of developing key infrastructure such as airports, seaports and telecommunication as the basis for industrial development. The telecommunication sector in many developing states, for example, is comparatively well developed and provides the backbone for a thriving telecommunication system, which is fundamentally critical for enhancing business development in all sectors of the economy.

Another key contributory factor in the success of the East Asian States is adequate economic and fiscal policies instituted by their governments. They are designed to support innovation and entrepreneurship. State-led banks and financial institutions have also given support to major industrial investment initiatives. Some small states have provided tax incentives for industrial development programmes, which have not resulted in the long-term benefits that were anticipated. These policies need to be reviewed in light of the going concerns for enclave type industries. Based on their industrial policies, many SIDS continue to support inward-looking industrial strategies. Recent events in world trade are beginning to bring awareness to policymakers about the realities and consequences of an import substitution industrialization policy regarding global competitiveness.

Without the political will to stimulate the private sector as an engine of growth, many SIDS cannot optimize their industrial potential. The state should adopt the role of initiator and facilitator of programmes in R&D, education, economic and technological restructuring. The Asian countries prioritized industrial development as the basis for economic success, while many other small states prioritize tourism and agriculture for their economic survival. It would be ludicrous to prescribe any set of generic blueprints for the development of policies in some jurisdictions because of various peculiarities across states.

The NIC countries bear historic and geographic similarities to many other small island developing states. All of them had a colonial past, possess little endowment in natural resources and had industrial policies that were fundamentally inward-looking. They initiated export-oriented policies and domestic utilisation of resources for the creation of an efficient modern infrastructure conducive to foreign investment. Most of them have undergone very rapid and rigorous economic restructuring. They have transitioned from export-oriented industrialization of traditional labour-intensive products to more complex and sophisticated export-oriented industrialization based on the export of capital and technology-intensive manufactured products.

Industrialization is a major force in structural change since it demands a shift from labour-intensive to more capital and technology-intensive activities. This policy is crucial to the future growth of developing states. The export-oriented industrialization policy can be fully operationalized by implementing all the instruments that would drive efficiency, productivity and competitiveness. Many regional groupings like those of the Caribbean, Africa and Asia and the Pacific, do not have the critical mass to attempt any mass production. Relatively high wages in labour-intensive manufacturing make it difficult for many small states to compete in an open market regime. It is therefore inevitable that competition should transcend regional boundaries into the international sphere. This approach will allow a country to expand production beyond domestic demand predicated on the efficiency and quality consciousness of enterprises through targeted international competitiveness. In so doing, the notion of targeting the domestic market is irrelevant to firms that are striving to become world-class and pursuing international competitiveness.

Structural and Economic Constraints

Economies often succeed when there is state intervention in investment in the promotion of export trade. Most governments in small developing states are unable to make any sizeable investments because several of these small states have become highly indebted and have been experiencing debt-servicing difficulties for many years. On the contrary, subsidizing a sector that grew rapidly does not necessarily imply that the growth should be attributed exclusively to the input of the government. The sector might have grown through other intervening factors. The governments of many small states have become constrained and vulnerable because of the pressure from international lending agencies including the IMF and the World Bank to restructure their economies and adopt more export-oriented policies.

Countries that pursue the path of export-led industrialization have experienced thriving economies in which the rate of savings is generally high and is often used to spur investment at a very rapid pace. Some studies have suggested that these saving rates are a result of very favourable economic factors such as high growth rates as well as in situations where consumption is lagging and incomes would increase. Small developing island states are not in a privileged position to enjoy these high saving rates that are necessary to bolster investments. To flourish industrially would require high levels of technologically skilled workers to develop the technical capacity that will put them in an advantageous position as they undertake primary processing. Low-level skills will narrowly define their exports, which will be concentrated on primary products in the form of processed or semi-processed products. Productivity growth, which is fundamentally critical to a developing country's success, cannot take place without industrialization and building technological capacity to trigger that growth.

In an era marked by rapid technological development, many developed countries including countries in Europe have quickly taken appropriate action to narrow the technology gap in ways that small island states are unable to do. It requires more than just buying technology. These developed economies have amassed accumulated capital at an impressive rate. The concept of capital accumulation is often broadened to include human capital.

Some East Asian countries encourage the transfer of technology from foreign investors and they have made enormous investments in human capital, educating large numbers of skilled engineers to enable them to absorb and adapt to the most advanced technology. Furthermore, they were amenable to accepting foreign investment and creating an economic atmosphere conducive to the growth of their enterprises. They have also combined these efforts with an emphasis on the most technologically advanced investments. Small states can ill afford to attract similar investors because the environment is not conducive to investment of the same scope.

Governments can play a critical role as a catalyst for growth without necessarily providing a lot of resources. This can be achieved through effective and strategic policy initiatives. The government's intervention is not exclusively about the quantum of subsidy or financial contribution provided, because in most small states there is a dearth of resources to make these interventions. More importantly, investments in human capital and infrastructure can increase the return on investment and thereby promote growth.

Unlike the private sector, most governments in developing states are not known to adapt quickly to changes and challenges to prevailing economic circumstances because of the bureaucratic processes that guide them. Most of the governments in East Asian countries are highly adaptive to circumstances that would redound to growth in the economy. They are known to learn from past mistakes and thereby adapt quickly in a manner that will compel them to make amends through new policy adoption. As their economies grow and become more sophisticated, the state quickly reviews their policies to encourage and promote higher levels of technology and higher value-added industries.

Enterprises in developed countries are at a distinct advantage because of the knowledge acquired from participating actively in international markets. Their success in export trade places them in an advantageous position where they can benefit from spillovers in terms of both marketing and production know-how. For example, success in producing intermediate goods requires knowledge of international standards which is driven by the demand for testing laboratories and quality and standards protocols. The relationship between quality and standards is reflected in the quality of products that these countries produce. Firms' involvement in international markets enable them to build the necessary networks and linkages with other firms operating in markets of similar interest. For example, they are predisposed to information that will allow them to develop databases on those firms that can supply information leading to advanced technology and training.

Lending institutions in small states are generally risk averse. They typically prefer to lend to firms operating in the domestic market than those that export abroad. In other words, these institutions operate in contradiction to the fact that success in exporting abroad is a better indicator of success than in selling domestically. The experience in small states suggests that banks are less informed about the intricacies of international trade, and therefore demonstrate less interest in foreign markets and the benefits of export. They consider it riskier to lend for projects with good export potential than for the domestic market as long as borrowing firms can repay their loans.

For many years, there have been several structural constraints that have precluded the governments of small developing states from making a more impactful contribution to firms' development in their quest to become successful in the export market. These issues can be exemplified by factors such as a lack of adequate and necessary infrastructure including roads, telecommunication, port facilities and preferential access to capital and foreign exchange. These shortcomings are exemplified in the lack of well developed capital markets and limited access to export markets due to the lack of quality exports that can gain entry into competitive markets. They are further hamstrung by the absence of licensing and other regulations designed to facilitate and promote a country's exports. These structural inefficiencies are serious constraints for the economic development of small states. The consequential challenge they face is to navigate the hurdles and find innovative ways around them.

Complementarity

Protectionist policies of many governments have supported non-competitive industries to avoid the necessary restructuring because of the avoidance of competition. A more serious problem is that declining industries often find it difficult to raise the necessary capital for restructuring programmes.

In this circumstance, financing institutions and investors are understandably reluctant to invest resources into enterprises that show little promise in short-term profitability, thereby leaving a huge gap that the government is unable to fill. As in the case of Singapore and other Asian states, one of the major factors that has contributed to their success is the development of a quality workforce. The economic success of many industrialised countries was predicated on the skills and education of their workers, which is universally accepted as fundamental pillars of a firm's success. Similarly, the productivity of small economies is dependent on these factors. Forfeiture of expenditure on human capital instead of investment in plant and machinery could be counterproductive.

Competition from industries that have an advantage in offering low wages, relatively cheap raw materials cost and easy migration of capital abroad, could have a crippling effect on those enterprises that engage in direct competition. For this reason, a progressive industrial policy should be designed to address these market failures. The imposition of tariffs, quotas and trade restrictions are often used as bailouts or stimulus for ailing businesses that have lost their competitive edge. These policies have had negative consequences on the rest of the economy. At the least, they often result in an increase in the price of inputs on which other industries depend for their survival.

The structural landscape of many industries is undergoing serious transformation driven by new digital technologies and business models. In light of these changing circumstances, the industrial sector in small states needs to be modernised and revolutionised to ensure that SMEs remain competitively relevant in the global markets in which they are competing. It is within this context that countries should embrace technological change, integrate products and services, develop technologies that use less energy, reduce waste, avoid pollution and invest in a workforce with the combination of the right skills.

Amidst the global transformative milieu, there is an apparent emphasis on services. The digital transition is evident in today's business environment and the associated models. This new paradigm shift will invariably affect every aspect of society and the economy in product offerings, marketing, finance, and generally, ways in which business is conducted. The pressure on natural resources is already leading to a more circular approach to manufacturing in that, from a technological viewpoint, the notion of waste is redefined to connote that every item of waste should be a resource for the next stage in the production process. Sustainable competitiveness will require industry to become leaner, greener and more digitalized, which will then transform traditional and new industries.

The current geopolitical realities have become more dynamic and as such, are having a profound effect on industry at all levels. Global competition, protectionism, market distortions, trade tensions and challenges to the rules-based system are all on the increase. New competitors are always emerging and are affecting the way business is conducted, including the imposition of new trading rules. Hence, barriers to competition are policies of the past. Competition engenders challenges that should be confronted head-on to survive in the market. It is therefore critical that the environment is established to enable and embolden entrepreneurs to transform their ideas into commercial ventures and for enterprises of all sizes to thrive and grow.

In the quest to promote industrial development, governments have a critical role to play in establishing policy direction. Industry has a leading role to play in the promotion of clean technology through the use of environmentally friendly raw materials and a commitment to reducing the carbon footprint of all countries. The traditional industrial model is being superseded by creativity and innovation, driven by digital platforms. New technologies are creating new products and business models that can be adapted for different markets thereby undermining the importance of location and size of firms. Increasingly interconnected economies and markets are facilitating the spread of technology and ideas around the world.

The knowledge economy is predicated on human intellectual capabilities and intangible assets rather than on physical inputs or natural resources. Science and technology is fundamentally based on all technical knowledge, whether novel or not. Scientific activities are mainly aimed at producing organised knowledge about physical, chemical and natural phenomena. The rapid expansion of knowledge and the increasing reliance on computerization, data analysis, creating and working with financial models and the ability to innovate are highly sought after in the modern economy. These factors are changing the economies of the developing countries to become more dependent on intellectual capital and skills, and less dependent on traditional production processes. The knowledge economy has placed the IT/ICT industries at the forefront of overall economic growth. It thrives on highly skilled workers and computer literacy which possesses expertise in algorithms and artificial intelligence (AI).

Digital innovation will continue to provide opportunities for businesses and individuals, particularly those in small states whose dependency on the domestic market is becoming less important because of expanding world trade. Small enterprises will now have an opportunity to access the global markets with their variety of products, notwithstanding the enormous challenges they face in pursuing such venture.

Large scale manufacturing industries still have an important role to play in the production of consumer goods. It is increasingly becoming evident that the new model of innovation in technology driven by creative talents will forge a new wave of industrial development in the future. Industrial change is accelerating at a seemingly unstoppable pace. It has revolutionised key sectors including retail and marketing, healthcare, energy, manufacturing, education and the creative industries.

The export-oriented industrialization model of industrial development has undoubtedly brought tremendous economic success to those countries that have applied it effectively. However, over the last couple of decades, this model has become exhausted. One of the main issues is that developing countries need to rid themselves of their reliance on "inconsistent capital inflows and commodity booms, which have often left them vulnerable to external shocks and prone to economic crises."

It has therefore become evident that small developing countries need a new model for economic growth and success. With this realization, some countries are beginning to de-industrialize by substituting manufacturing for services at much lower levels of income. Consequently, employment in manufacturing as a share of total employment has fallen dramatically. One notable case in point is Britain, which gradually lost its competitiveness in manufacturing after the Second World War. The situation worsened after the 1980s and has led to its rapid de-industrialization.

The services sector has shown tremendous growth over the past few decades. It has played a significant role in the world economy contributing to about 49% of global employment. The biggest contributors to the recent economic growth have been knowledge-intensive business services such as telecommunications, computer and IT services, R&D services, financial and legal services, accountancy, management consultancy services, architecture, engineering and other technical and professional services.

Trade-in services enabled by information and communication technologies are becoming an engine of growth for many developing countries looking to diversify from commodities-based economies. For this approach to be successful, governments should establish effective policies and develop the appropriate regulatory framework, infrastructure and human capital. The application of information technology is radically changing the way services are being packaged and used. There is now a constant quest in the services sector to segmentize any business function in which knowledge can be commoditized and packaged as a "product," where ownership can be established, production can be scaled up, and trade can take place separately from production. This innovative business approach will be transformative to SMEs in their quest to create globally integrated services firms.

As many developing countries begin to deindustrialize and adopt new models of development that are perceived as 'game changers,' such models should incorporate support for innovation and the development of new knowledge. Labour-intensive industries are being replaced by high technology machinery and trade in services, which will provide a greater level of efficiency and productivity. When labour-intensive industries begin to lose their competitive advantage in world markets, it is an indication of the need for industry to begin to restructure to arrest the decline. Such decline can result in real social costs and the loss of jobs for many people. To obtain other jobs may require those in search of a job to relocate or learn new skills, which will necessitate adjustments that are sometimes slow and costly with a lasting effect on the economy.

Services, unlike goods, are not subject to economies of scale in the same manner. For those countries in the lower income bracket, services may provide a platform for "leap-frogging" stages of development if the government can help to create conditions necessary for developing the services sector. Governments can 'create' comparative advantage in service delivery since services are all about people and human capital rather than physical capital. The role of policy can be a determinant in this regard. Services exports are making a higher contribution to GDP in poorer countries than in middle-income economies; just under 6% for lower middle-income economies and low income economies.

It is not being suggested or implied that services should replace manufacturing. The ideal situation is for both to strive simultaneously and productively. It is critically important to a country's development that sectors coexist as viable entities. A strong services sector complements a thriving manufacturing sector. The important challenge is to address the structural and policy shortcomings that are obstacles to efficient manufacturing and service sectors.

In recent years, ICT and other new technologies have created new possibilities in the service sector, particularly by improving their tradeability. It is therefore strategically critical to create a good balance between services and manufacturing to maximise growth potential. The services sector ideally complements the manufacturing sector more so in helping to increase productivity in manufacturing as well as other sectors of the economy. This is particularly applicable in business services as they provide inputs such as finance, legal services, human resources, marketing and ICT. To maximise cost-efficiency, the manufacturing sector can outsource these services rather than create specialised internal departments to execute these functions.

Chapter 14

Industrial Policy Issues; Perspectives and Proposals

It is incumbent on national policymakers to design comprehensive industrial policies cognisant of the importance of the need to create an environment that enables the development of diverse industries and in particular, one that will reenergize the growth of the manufacturing sector in SIDS. As was discussed earlier, the industrial sector in general, and particularly manufacturing, holds tremendous potential for job creation, foreign exchange earnings and the avoidance of the scourge of poverty.

    An industrial policy delineates the roles of the public and private sectors, joint and cooperative sectors, as well as small, medium, and large-scale industries. It outlines the government's policy towards industries, foreign capital and technology, labour and tariffs. There are contradictory viewpoints regarding the relevance of industrial policies. One school of thought is that they are not necessarily relevant in an age of globalization where the global market is dictated by several policy-related guidelines which tend to supersede those that are national and regional in scope.

    Conversely, some argue that industrial policies are still relevant particularly for small developing states since they are generally geared towards identifying and harmonizing policy issues and goals.

Furthermore, the aim of an industrial policy is to improve market conditions and facilitate sustainable production at the regional and national level. Industrial policies should be well structured and targeted to potential areas of growth. The success of policies depends to a large extent on what lessons of the past can be incorporated into the decision making process at the macro level. It is important to consider how decisions will affect corporate planning and those designed to avoid mistakes that have beset the sector including export subsidies, picking of winners and high tariff protection for infant industries.

An industrial policy cannot be separate and distinct from a development strategy of a country or a region which has a common developmental agenda. In other words, there should be an integrated approach to industrial development that would catalyze the productive activities within the sector, enhance growth, increase employment and improve standard of living. The classical arguments relating to market failure, which justify the argument for industrial policies surround the lack of markets in some cases, weak institutional structures and lack of capacity and resources. Small manufacturing enterprises have been experiencing enormous challenges in areas such as financing, working capital and infrastructure.

These issues cannot be addressed satisfactorily unless there is government's intervention driven by appropriate industrial policies. A national industrial policy should be designed with the overall goal of increasing a country's production capacity and thereby the export potential of selected products as well as increasing the competitiveness among major industries. Hence, an industrial policy should have the following desirable objectives:

(i)   Facilitate, foster and maintain sustained growth in productivity;
(ii) Provide opportunities for gainful employment;
(iii) Achieve optimal utilisation of resources;
(iv) Enable manufacturing enterprises to attain world-class status;
(v)   Attain the level of international competitiveness.

Considering the discussion in question, an Industrial Policy can be defined as a stated set of actions or interventions to be pursued by the state to affect the way factors of production are being distributed across national industries. In shaping an industrial policy, cognisance should be taken of the overall competitiveness of the industrial landscape *vis à vis* many of the issues discussed earlier including technology transfer mechanisms, labour cost, the productivity gap and the impact of past and existing policies and current market dynamics.

The traditional framework for the analysis of an industrial policy is based on the notion that government interventions may be desirable in the face of significant market failures. Where markets lead to misallocation of resources, corrective government action can sometimes improve economic efficiency. Market failures may also arise when competitive markets either do not exist or are incomplete. Indeed, market failure is not seen as a sufficient condition for intervention, but as a screening criterion to identify areas where such interventions may be beneficial and where further analysis of feasible policy options would typically be justified. In such circumstances, the role of government's intervention is to correct such failure and to put the market on a freely competitive trajectory.

The rationale for the development of an industrial policy is based on the notion that market forces by themselves without the support of the state, cannot be sufficient to bring about sustained development of the productive base of the economy. Hence, an industrial policy should be designed to ensure that there is an integration of sectors and cross-sectoral linkages so that the benefits can be augmented. Standards and quality, financing, human resource development, environmental sustainability and the like, that have beset the industrial sector for many years, will be addressed wholesomely giving effect to a more dynamic, efficient and competitive industrial sector. Industrial policies also provide the opportunity for governments to collaboratively tackle the vexing issue of sea and air transport among others in a coordinated manner. These logistical issues are important to the success of the sector in getting products to the market at the right time in the right quantity without any hassle.

Given the foregoing, it is necessary to put forward recommendations that will attract the attention of national policymakers who should design comprehensive industrial policies taking into consideration the importance of creating an environment that enables the development of diverse industries. Such policies should also energize the growth of the manufacturing sector by enabling enterprises to grow and flourish through clear policy guidelines. It should also address areas such as energy cost for industrial purposes, environmental protection, technology transfer, and effective incentive schemes.

A region which is desirous of pursuing economic integration should first define its own path to economic success. Realistically, several policy interventions may not be harmonized at the regional level, notwithstanding the similarities shared by several states, since they are at varying stages of development with different national priority issues and agendas. Despite the foregoing, the onus is on government at the national and regional level to find the political will and common ground on which to collaborate to achieve optimisation of resources and economies of scale. This will enable states to achieve greater synergies by providing the environment and the impetus to propel growth in industrial development.

Industrial Policy Structure and Framework

An industrial policy can have either a demand-side or a supply-side orientation. The concept of an industrial policy framework oriented towards the demand side of the market refers to the practice of promoting specific industries by providing a "captive market" to producers through tariffs, quotas and other measures. This framework which is focused only on fostering domestic demand is usually developed to protect an infant industry or vulnerable industries within the economy, which are an offshoot of the import substitution industrialization policy of the past. Some states have a long history of sector-specific protection, especially in the agriculture and manufacturing sectors. One must be mindful that this approach may be incongruous with the WTO's goal of promoting liberalised trade across the globe.

This practice has been frowned upon by the World Trade Organization, which has endeavoured to minimise the use of this type of demand-oriented trade policy. Even within countries, there is often much debate surrounding whether this approach does indeed lead to the development of the protected sectors. Nonetheless, no country has evolved without some form of demand-driven industrial policy, to this day, countries still provide some form of protection to vulnerable industries.

Supply-side oriented policies, on the other hand, refer to policies aimed at boosting the production capacity of enterprises in the areas of investment incentives, human resource development, support for R&D and provision of information on production methods. Policymakers seem to have a clear preference for supply-side oriented industrial policy, because this policy option has yielded some success after many of the obstacles that have impeded the growth of small firms have been overcome. Among the obstacles are inadequate financial institutions, underdeveloped human capital and challenges regarding technology and marketing.

The shift towards supply-oriented policies had been witnessed at the international level. Nevertheless, it is still unclear if this is the direction in which most governments are moving, even with the gradual elimination of some forms of protection. Some governments have not fully moved away from demand-oriented policies, despite the use of trade incentives which is somewhat on the decline.

More recent industrial policies encapsulate export promotion, output growth and investment, higher level of productivity, promotion of innovation and competitiveness. These features are articulated in the various policy interventions designed to correct failures and challenges within the sector. The overall goal of a national industrial policy should be to increase a country's production capacity and thereby the export potential of selected products, which will ultimately increase the competitiveness among major industries. It should target two major development objectives namely, poverty alleviation and restructuring of the economy. The right model for industrial policy is not that of an autonomous government applying taxes or subsidies, but of strategic collaboration between the private sector and the government to uncover where the most significant obstacles to restructuring lie, and what type of interventions are most likely to remove them.

In line with these development objectives, an industrial strategy should therefore aim at a competitive industrialization programme that would generate employment led by export-oriented industrialization. These programmes should be geared at promoting greater participation of the general population in the industrial development process of the country. This strategy should be aggressively pursued by establishing a supportive industrial and institutional infrastructure. Over the past couple of decades, ACP countries have depended heavily on agriculture-based activities. Agricultural development was predicated on the availability of large estates particularly located in rural communities; the availability of relatively cheap labour and the notion that their export markets would always provide some form of protective regime for their products.

The option for economic diversity was constrained by inward-looking economic policies, a poorly educated workforce or in more recent times a workforce inappropriately trained for the development needs of the country and underdeveloped and in some cases non-existent infrastructure to support industrial development. Protectionist policies of former colonial powers in support of export trade from the territories they once governed, hardly provided any incentive for the development of the industrial sector. Succeeding years have engendered an unprecedented paradigm shift in the international market. This has forced some states to reassess their position in respect of what was happening globally and the challenges they face. It was evident that many developing countries had to radically overhaul their economic and industrial policies to address the existential challenges of the global market.

Over many decades, the pattern in the agriculture sector, which had been the main economic driver in many small states was that, one main export crop successively gave way to another as market demand and trade conditions changed. Over the last couple of years, the design of industrial policy has significantly changed because of new rules governing international trade, including the rise of global value chains, marketing networks, and other aspects of globalization. Within the last several decades, there has been considerable change in market conditions because of liberalised trade dictated by the World Trade Organization (WTO). This has resulted in the removal of major trade protection regimes and the establishment of new protocols for some commodities.

The impact of such decisions has created enormous challenges for those small states. Other traditional industries such as sugar, rum and rice have struggled to survive in the face of aggressive and dynamic global competition. All these events took place in the absence of a strategy to negotiate the orderly phase-out of market preferences. Furthermore, changes were effected without the appropriate technical and financial assistance to help in the transition, including taking care of the displaced labour by providing alternative livelihood for the citizens of the small states.

Consequently, regional governments had to reprioritize and develop new strategies to support industrial activities and strengthen those that will now form the engine of economic growth for the future. In this context, new policy framework should be designed within which timely interventions can be made to strengthen existing industries and encourage new developments in the sector, taking advantage of regional production and available resources.

When developing states form themselves into regional groupings such as the Caribbean Community (CARICOM), Pacific Islands Forum and African Small Island Developing States, it is in their interest to devise harmonised regional policies apart from their individual industrial policies. These individual policies which invariably reflect national interest should feed into a broader regional policy. Such arrangements should be made in a manner that conveys the philosophy that functional cooperation on issues of mutual interest would be beneficial to all member states. The converse would have deleterious effects if member states pursued their individual agendas in a global environment that does not recognise the importance of small and microstates. Harmonised policies should be designed to promote the following ideals:

- Promote cross border employment of natural resources, human resources, capital, technology and management capabilities for the production of goods and services on a sustainable basis;
- Establish linkages among economic sectors and enterprises within a particular grouping;
- Develop regional economic enterprises capable of achieving scales of production to enhance the competitiveness of industries in domestic and extra-regional markets;
- Establish a viable micro and small and medium size enterprise sector;
- Enhance and diversify the production of goods and services for both the export and domestic markets;
- Facilitate public and private sector collaboration to secure market-led production of goods and services;
- Enhance industrial production on an environmentally sustainable basis;
- Establish a stable industrial climate that would promote industrial relations.

It is incumbent on the government to engage the private sector in meaningful discussions towards developing a relevant industrial policy that will be aligned to the country's strategic development. Such policy should consider the country's resource endowment, the promotion of direct foreign investment and strengthening the technology transfer capability. It is not reasonable to expect governments to give fiscal support for development policies that are outdated and irrelevant. The import substitution policy for many years has protected the local industries and it has not yet been determined to what extent the country has benefited from it.

In a globalized environment with a liberalized trade regime, such policy has outlived its usefulness. There is a need for policies that support an outward-looking economy, which will enable the country to develop and utilise skills in new technologies that will promote, for instance, export industrialization.

Macroeconomic Policies

Industrial policy initiatives should take cognisance of the macroeconomic policies and the existential economic circumstance of small states. Such policy goals should include among other things:

1. Sustained economic growth;
2. High employment;
3. Stable price (low inflation);
4. Rising living standards;
5. High productivity.

In the 1980s and 1990s, several countries undertook structural adjustments and reforms to make their economic environments more open and competitive. Among those changes that resulted was the shift from import substitution to export promotion industrialization, trade and foreign exchange liberalisation, the dismantling of quotas, and the implementation of a harmonized lower tariff. An example is, the lower tariff rates such as those implemented under the (Common External Tariff of CARICOM) could benefit regional manufacturers by reducing the price of imported raw materials, manufacturing plants and other factor inputs. Furthermore, in some countries, many state monopolies that have burdened government finances for a long time have now been privatized.

A sound exchange rate policy is crucial to strengthening manufacturing export competitiveness. It is recognised that unstable and overvalued exchange rates have in the past, retarded the competitiveness of a country's exports. Overvalued exchange rates have artificially inflated the price of exports and deflated the price of competing imports, resulting in reduced demand for price-elastic exports and increased demand for imports. An overvalued exchange rate is essentially a tax on exports and a subsidy for imports. Therefore, to help exporters provide competitively priced products, policymakers should ensure a realistic exchange rate. This is particularly important for a markets where most of its manufacturers are not well differentiated, making price competitiveness more important.

Most industrial policies today are subordinated to tax, tariffs and trade rules of the WTO and the General Agreements of Tariffs and Trade. It is important to create an enabling stable macro-economic environment through appropriate monetary, fiscal and other policies consistent with industrial development targets. Governments' intervention through its fiscal, monetary, and supply-side policies should be geared towards stimulating the market and making the investment and production climate more enabling. These policies are designed to influence taxation and promote the government's spending and borrowing which would impact industrial activities relating to investment and government spending. Supply-side policies are designed to make industries more efficient and thereby stimulate the growth of national and regional output.

In the context of the foregoing, the measures discussed to address the issues of efficiency and competitiveness in the global market should be supported by appropriate and timely government interventions. It is, therefore, necessary and desirable for governments of SIDS to develop industrial policies that will chart the growth path through competitiveness for these enterprises. In conjunction with clear policy direction and the adoption of appropriate development models, policies should address areas such as energy cost for industrial purposes, fiscal interventions where necessary, environmental protection and technology transfer, among others.

The enactment of the appropriate fiscal and legislative measures is necessary to establish the framework to promote foreign direct investment, as well as other measures to bolster manufacturing in small developing states, inclusive of strengthening state supporting institutions such as development banks, standards bodies, development corporations and the like.

Sectoral Policy

Several aspects of industrial development would invariably impact or have some relationship with other sectors of the economy. Hence, an industrial policy needs to be cross-sectoral in scope and provide the contextual framework for synergies and complementarity. They must therefore address issues such as intellectual property rights, competitiveness, energy and the environment, information and communication technology (ICT), technology transfer, standards and quality.

Policy initiatives should address the issue of infrastructure development, which will provide the framework for aligning sectoral policies, designing national development plans, harmonising regulatory regimes and investment codes, attracting seed capital, and mobilizing investment resources. It must also address initiatives to integrate transport, communications energy infrastructure and services across regions. Such policy initiatives should focus primarily on reducing the cost of conducting business within a specific region and facilitating cross border mobility, which will contribute to increased investment and competitiveness among industries.

An industry policy should delineate the need for sustainable infrastructure systems that meet economic demand and provide basic services for small and medium enterprises (SMEs) development. These infrastructure systems should be reliable, efficient, affordable and environmentally sound and should provide the opportunity for growth that will enable all enterprises to compete in regional and international markets. To meet these requirements, policies must also outline the appropriate regulatory framework for monitoring performance and liberalising access to infrastructure service markets.

The natural endowment and peculiarities of small states should be recognised when crafting an industrial policy. Moreover, provisions should be made for cross border cooperation and integration. The policy should be designed in such a way to facilitate the establishment of linkages to important sectors like agriculture, services, and energy. It should also provide the framework for state intervention, particularly in infrastructure development to address extant concerns and facilitate growth in the sector.

Traditional Approach to Industrial Policy

Traditionally, industrial policy positions have been accentuated in position papers (white paper, green papers, and the like). Several sectoral studies have been conducted in selected areas to determine the competitive status of selected commodities. In the past, industrial policies were dictated primarily by governments with little input from the private sector. Most policy positions have largely been reactive, inward-looking (driven by import substitution goals), bureaucratic and demand-side based.

The introduction of new policy measures has given rise to little institutional change except for increasing the regulatory capacity. Issues such as capacity building and institutional strengthening were ignored for the most part. This was because such policy changes have been primarily based on price interventions through taxes, import tariffs or subsidies. A modern industrial policy requires pro-active intervention on key developmental issues including adding value to primary products as well as relevant changes for its implementation. Changes that support institutional repositioning and strengthening should reflect new policy direction. Policies require continuous monitoring of the operating environment to identify and address policy bottlenecks that would stymie its effectiveness. Inhibiting factors and constraints within the framework of the implementation of the policy should be identified and removed to achieve sustainable development.

ISI Policy and Strategy Consideration

It was mentioned earlier that the import substitution industrialization (ISI) strategy in many small states has failed to meet its objectives. Although it has led to the development of a fledgling manufacturing sector in some countries, it has not, however, met expectations in terms of import replacement, employment, and contribution to the growth of the economy. It has been argued that import substitution, which protects the domestic industry from foreign competition has encouraged inefficiency, low-quality production and a lack of competitive vision.

Logically, an industrial strategy should inform an industrial policy. The strategy requires a creative and innovative approach and close public sector and private sector partnership and participation. The overall goal of the strategy should focus on enhancing the productivity and efficiency of manufacturing firms, diversifying export products and markets, and identifying new markets for niche products with a higher processing and value-added component than the commodities currently being exported. It should consist of the following elements:

(i) prioritisation of key industries
(ii) (ii) budgetary allocation to foster rapid manufacturing growth;
(iii) investment to support physical infrastructure and allied industries
(iv) improvement of macroeconomic, business and legal environment;
(v) emphasis on private sector-led growth, export-oriented manufacturing production and foreign direct investment and any other policies that will provide the impetus for sustainable growth in export trade.

Many of the products that were produced under the influence of this ISI strategy have failed to meet international market standards. Furthermore, investment was probably misdirected into areas in which there is no comparative advantage, while activities with possible comparative advantage were neglected. Import substitution industrialization has encouraged high import dependence because of heavy dependence on imported inputs of machinery and raw materials. This has led to the subsidization of ISI by the foreign exchange earning export sectors.

It was mentioned earlier that most of the economies of small states had been traditionally inward-looking. Most of these policies, albeit implicit, promoted import substitution, where domestic manufacturing industries were protected using high tariff rates and other qualitative restrictions to limit or exclude competition from imports resulting from the advent of free trade. WTO rules have hitherto restricted such protective policy. Generally, import substitution industrialization strategy has produced disappointing results for many developing countries.

The tenet of a progressive strategy should be geared towards the development of a competitive and productive set of activities that will foster and promote growth within enterprises with the long term goal of increase productivity, enhance competitiveness and increase employment within a sustainable framework. The strategy should point to the justification for and the conduct of industrial policy to address existential issues in manufacturing enterprises by ensuring that they are equipped with the necessary tools to compete in the international market. There should be adequate incentives to encourage growth within the sectors by ameliorating the challenges that retard the growth of firms. Strategies for picking winners should be avoided, but this does not preclude the use of selective and targeted financial and technical assistance to some sectors with high growth potential.

Industrial Policy Priorities for Small Economies

The political independence of many former colonial states in the 1960s and 1970s paved the way for economic independence throughout the Caribbean and other regions. Most of these small developing states have used industrial policies to stimulate growth in industry. In the past, former colonies of Britain, France and Holland were offered preferential access to traditional markets in Europe for their agricultural products. This approach allowed newly independent colonial states time to invest in new machinery and equipment and develop new industries. Independence brought new levels of aspiration in all sectors of national development, not least in industry. The development of institutions like banking, insurance, hotels and the like, became priority. Economic development and poverty reduction were therefore contingent on the success of these institutions.

To accelerate the growth in industry, countries established Export Processing Zones (EPZ) to which manufacturers were attracted because of relatively low labour costs, exemption from taxes on profits as well as purchases of imported inputs with little or no foreign exchange or government restrictions. The EPZ is no longer a key attraction for FDI since there are other economic and technological factors that could attract foreign direct investment. With the advent of free trade and a dynamic global market driven by innovation and competitiveness, the EPZ is gradually becoming less attractive as a model to promote investment among firms. Small states should therefore devise other policy interventions that will promote the growth of small enterprises.

Strategic Goals of an Effective Industrial Policy

It can be argued that industrial policy is still relevant today even in a market-driven era of globalization. However, based on the lessons learnt from the mistakes of the ISI period, an industrial policy should be aligned to meet the demands of a competitive global market. No longer can incentives be provided to protect inefficient producers. Hence, new policies that promote efficiency, quality and productivity are required to drive competitiveness. Generally, financial assistance and incentives are useful in providing support to facilitate the development of activities in manufacturing and services, if well managed. Strategies for picking winners should be avoided, but this does not preclude the use of selective and targeted financial and technical assistance to some sectors.

An industrial policy should also be integrated within the overall development strategy of the country, and it should reflect the collective goals of a particular region to which countries are aligned. The following strategic goals are designed to be pursued by a harmonized policy framework in the context of a regional grouping of small states. It is anticipated that developing states will commit themselves to the ideals of a strong industrial development programme, geared towards improving the industrial sector with the following goals that are envisaged to attract investment and inspire growth within the targeted sectors:

(i) Improvement of the physical infrastructure through public and private sector partnership in infrastructure development such as roads, telecommunication, adequate water and power supply, etc. will facilitate accelerated growth in industrial development.

(ii) Restructure and strengthen industrial parks and allied facilities so that firms that operate within these facilities can provide quality services to their clients through enhanced collaboration among entities. This will increase the probability of success among firms and render them less likely to struggle in the face of the vagaries of the competitive global environment.

(iii) Establishment of an effective system of "single point contact" and "time-bound clearance" to expedite new business development by eliminating tedious bureaucratic processes.

(iv) Promotion of quality and standards by developing the appropriate systems that will provide the framework for competitiveness among regional industries.

(v) Provision of specialized incentive schemes for SMEs and leveraging of funds that will be available with predetermined conditions through appropriate institutions.

(vi) Creation of enabling business climate and the provision of targeted incentives for selected industrial entities for technology upgradation and modernization to enable them to face the challenges of global competition.

(vii) Building a culture of entrepreneurship through appropriate programmes in the school curriculum and national skills training and mentorship programmes for young people in targeted areas.

(viii) Development of an effective transportation system that will facilitate the movement of goods and services throughout the region and the international market in the most efficient, efficacious, and timely manner.

(ix) Promotion of world-class manufacturing techniques within manufacturing enterprises as effective tools to promote efficiency and quality and to enhance productivity.

Sectoral Policy Shift towards Sustainable Growth

The line of demarcation between industrial policy and corporate policy is becoming progressively blurred. The manufacturing sector in some small states is dominated by a handful of powerful companies that influence policy directions and which collectively can have a significant impact on the sector. At the company level, strategies including mergers and acquisitions, trade disputes, the emergence of company networks, outsourcing, relocation, restructuring; disputes on state aid and anti-dumping procedures can influence policy direction in favour of powerful companies. Some companies can also influence changes in the regulatory framework such as merger and take-over directives, competition laws, company taxation, corporate governance and framework for intellectual property rights. Many of these issues were flagged and debated during discussions about operational conditions in the telecommunication market in the Caribbean with the merger or acquisition of some service providers who were desirous of continuing their monopolistic dominance in the market.

The global trend in maintaining sustained growth is to make incremental improvements in factors of production. This would mean that drivers of growth should be identified and prioritized. In traditional industries such as sugar, rum, rice and bananas, there needs to be some level of diversification. Although there have been some diversification efforts in some traditional sectors, they have occurred in the form of niche manufacturing and diversification from traditional industries to address market demands. The pace of such diversification should be accelerated to reverse further decline in these sectors.

The prospects for sustained growth are especially challenging for small states with competing scarce resources. The gradual migration of citizens to developed countries, seeking employment opportunities, reduces the availability of skilled professionals in the domestic market. Public sector investment should be carefully planned and executed promptly. Many small island states are disaster-prone, and over the last several decades, there have been a series of major disasters including floods and hurricanes that have significantly impacted the economies of several small states including those of the Caribbean, thereby affecting their GDP. The impact of these events has also affected capital accumulation and exacerbated the output volatility that small countries experience.

The reduction in foreign aid over the last couple decades and the withdrawal of concessions by multinational donors means that the islands will need to increase their foreign direct investment (FDI) as an alternative source of capital and investment. While some Caribbean states for instance, have attempted to shift towards a private sector-led development strategy, high levels of government debt have crowded out private investment, and in some cases, public investment as well. This has led to high levels of taxation which further exacerbate the fiscal problem. Hence, efficient private domestic investment as well as FDI will be key to sustained future growth.

The Caribbean share of world FDI has been declining since the 1990s, partly due to strong competition for foreign investment from Latin America. Furthermore, foreign investment policies that have historically been restrictive, have become liberal but are still ad-hoc. The challenge for small states is to maintain and increase private investment rates in the face of erosion in preferences. This will otherwise lead to a decline in returns on investment in protected segments and increasing international competition. If developing countries are to close the technology gap with high-income countries, upgrade managerial skills and develop their export market, they will need to position themselves to attract FDI flows. This can only be done by devising attractive and effective FDI policies.

Attracting productive private investment requires that government provides basic social services efficiently and shape other aspects of the investment climate such as macroeconomic and political stability, a sound regulatory framework and adequate physical and social infrastructure. A favourable investment climate will encourage investments to be channelled into productive sectors and will also encourage investment in non-traditional areas. A good investment climate unleashes competitive forces that stimulate innovation and productivity. If small developing states are desirous of make meaningful contributions towards enhancing living standards via world trade, then they should undertake measures that would allow them to transform their production bases to produce higher value-added manufacturing goods and services.

Evidence suggests that in most small states, the public sector is the main employer of labour because a less dominant private sector cannot be relied on to generate the levels of employment to absorb surplus labour. The leading role of the public sector in terms of employment is evident in the high levels of expenditure on employment and social services, thereby increasing the fiscal burden of the state. In the absence of a well-developed private sector and weak to nonexistent civil-society institutions, citizens are forced to look towards the state for solutions to society's domestic and social ills, as well as the provision of life's necessities. Consequently, some small states tend to rack up higher government spending per capita on goods and services, wages and salaries and capital investment, thereby imposing a heavy fiscal burden on the state.

Policy Constraints within the SME Sector

The Caribbean Community (CARICOM) in recognition of the importance of the industrial sector in advancing the region's thrust towards a single market and economy, commissioned a consultation/study of the industrial sector which culminated in an industrial policy for the region's fifteen member states. Indeed, the implementation of such a policy poses its unique set of challenges in a region with autonomous individual island states with their national programmes geared at private sector development. Though not impossible, the achievement of this noble goal will require the coordination of national industrial policies, the establishment of an investor-friendly environment which speaks to several cross-cutting issues, and the strengthening of the institutional framework necessary to support the growth of the industrial sector.

The process involved a series of organised consultations with member states based on several stated objectives. The issues that were discussed during the consultations and which formed the framework for policy interventions were all thematic priorities that are important to the member states concerned. The initiatives proposed are considered important in reviving the prospects of an ailing industrial sector in the region. Foremost among them are creativity and innovation in product offering and processes, human resource development to improve technical and entrepreneurial skills and standards and quality. Most of the initiatives proffered are not necessarily new but have previously been done in a rather piecemeal and ineffective manner in some sub-sectors to meet the stated objectives.

The manufacturing sector and more specifically the SME sub-sector in the region remains underdeveloped and largely without significant institutional, technical and financial support. Over many years, several leading institutions including the World Bank, the OAS and UNIDO have funded numerous studies geared towards understanding the sector vis á vis some of the underlining challenges affecting its growth and performance. The results have been unanimous and well documented that some of the constraints that are plaguing the manufacturing sector have limited its ability to engage in meaningful value-added activities to produce innovative products for a dynamic global market. The following factors have been identified as constraints that have retarded the growth of small enterprises in the region for many years.

(i) Inconsistent and insufficient supply of raw material;
(ii) Poor quality raw material supply and high losses during transport from farm to factory;
(iii) Inappropriate or inefficient technologies;
(iv) Poorly trained personnel and lack of qualified technical staff;
(v) Lack of managerial and entrepreneurial skills;
(vi) Lack of finance and access to credit.

The policy construct was hypothesized on the notion that if the constraints within the industrial sector can be addressed efficaciously it will redound to increased wealth creation and employment throughout the region. In so doing, emphasis should be placed on building the relevant capacity within the sector. In the medium to long term, the state and the private sector should as a matter of priority invest in the development or upgrade of research facilities. The maximization and utilisation of existing facilities as well as using the most appropriate and current technology available is also critical for the sector to develop.

There is unanimity in the conclusion from numerous studies conducted on the status of some regions' SMEs sub-sector, which expressly states that SMEs are struggling to cope with the global market realities and the vagaries emanating from them. For the most part, the lack of resources and technical training in several critical areas have impacted negatively on the region's competitiveness in the market in which they operate and add to their vulnerability and eventual decline. At the macro level, there are hardly any effective policies in place to support and strengthen the survival of this sector. SMEs are often regarded as loose and disorganized without a collective voice to articulate their concerns.

Many small enterprises are devoid of working capital and do not have access to credit, which will allow them to acquire the financial resources to invest in new processes, improve the quality of the human resource and pursue technological innovation. Without the latter, there could hardly be any long-term strategic vision that speaks to innovativeness and creativity. There are also exogenous factors over which SMEs have little or no control. It is therefore imperative to design a system which is flexible and can respond to external shocks and thereby improve their development performance and promote industrial development as a key driver to economic growth. Governments will need to embrace evidence-based, participatory and ambitious, yet realistic industrial policies and strategies. This will require coordinated actions in various areas of the economy, including the establishment of forward and backward linkages among agriculture, manufacturing and other key sectors. The following decisions are critical in strengthening the industrial sector:

(i) Investments in infrastructure development for energy production;

(ii) Improvements in the transport and telecommunication infrastructure and services;
(iii) Removal of bottlenecks to improve business efficiency by promoting good governance using the court system to curb corruption;
(iv) Investments in technical and vocational education and entrepreneurial development;
(v) The creation of a science-based industrial sector with linkage to industry, research institutions and financial actors.

In more progressive enterprises where the converse is true, SMEs have strived and become competitive entities. They have been known to contribute significantly (particularly in smaller economies) to economic development. Strong SMEs have created employment and have positively addressed the issue of poverty among the rural poor and vulnerable groups such as women, youth and the less educated. As a result of the contribution of this sector, there ought to be realistic policy intervention for SMEs at the state and regional levels.

Human Resource and Industry-led Growth

Undoubtedly, education is the basis for change in every society, and it is a vital factor in the management of the factors of production and most firms are willing to pay a substantially higher wage for qualified workers. Education for basic needs supersedes traditional education since basic education is the foundation on which specialised skills can be developed. The benefit of education and training is universally recognised. Hence, it should be prioritized and developed in a manner that will increase productivity and allow citizens to participate meaningfully in community life. Special attention should be paid to human resource planning and development so that education and training programmes can be designed to respond to the changing needs of society.

The lack of capacity in enterprises in several functional areas can be attributed to a lack of appropriate training and the inability of small enterprises to adequately remunerate skilled professionals. It has been widely recognised that this sector has been reputed for low productivity, inefficiency and waste, primarily on the basis that it has only been able to attract personnel with limited and low-level skills in critical areas. Labour is the most important input into any manufacturing and production system and should therefore be treated as such. A country needs to identify programmes geared towards developing the relevant skillset to address the needs of the industrial sector and streamline its students to pursue those programmes.

Education and training cannot be done in a vacuum and in isolation of wider development goals. It is within this context that linkages between universities and the private sector should be created so that the mix of skillset that are required to drive important sectors like manufacturing, agriculture and the services sectors are provided by institutions with the mandate and resources to do so. The private sector is always in need of specialised skills to enhance its ability to increase productivity and allow for diversification into new areas of business. Therefore, firms are willing to pay substantially higher wages for well educated and trained workers. The nexus between industry and relevant post-secondary institutions needs to be strong to influence curriculum development in the programmes that are being offered at the post-secondary level. Undoubtedly, education is the impetus for change in any society and is also a critical factor in production output.

Training of the workforce is a necessary ingredient for the growth of firms. Improvement in the competencies and knowledge of the entire staff is a critical undertaking for the firm's success. Basic education is an important area that should be emphasised as an intellectual investment in the form of higher education, which is a policy matter deserving of being given priority. In the long run, it is undeniably the best way to develop the requisite skills and knowledge to enable firms to respond to the needs of the industrial sector. Education is not limited to school and university; it also applies to working life. Continuing training or lifelong learning for employees enables firms to become progressive and relevant to the needs of the market and society in general.

Basic education should promote the development of technical

skills that are necessary for the operation and repairs of equipment and machinery as a function of ways to increase productivity in small scale enterprises. The cost of human resource development is exorbitant, and the cost continues to escalate. For this reason, the need to prioritize and optimize opportunities for education and training should be appreciated and reflected in strategic planning as a means of maximizing the country's scarce resources. Attention should therefore be paid to human resource planning and development so that training programmes can be geared to respond to the changing needs of society that will lead to the creation of wealth.

At the national level, the state has a responsibility to respond to the needs of both the private and public sectors by churning out a cadre of students from its institutions with the fundamental capacity for advanced training in science and technology, management and vocational disciplines. The school curriculum should be geared to ensure that all educational institutions in their subject offerings are responsive to national and to some extent regional development needs. This may include revamping the school curriculum to reflect the realities of a country's development path with a focus on the skills required in key sectors of the economy by key sectors of development.

Small states need to build an entrepreneurial culture within their citizenry. Developing an entrepreneurship culture is a central pillar of an industrial policy and the functioning of market economies. Entrepreneurs keep the wheel of industry spinning because of their ability to realise their objectives by exploring and exploiting business opportunities in a market environment and taking risks in investing in potentially profitable ventures. While they may not enjoy success in every venture, a country with a substantial entrepreneurial spirit is likely to consistently generate new products and services to keep the market buoyant.

In many small developing states, institutions are beginning to offer programmes in "Entrepreneurship" targeting students with the aptitude to develop and manage their own businesses. Such programmes will help to sensitise students and provide them with an appreciation of the subject matter. Entrepreneurship should not only be reserved for students with a proclivity towards business ventures, but students of science have also been noted for their success in business. An exemplary story is that of Filtronic plc, a British multinational electronic company which has become world-renowned with expertise in commercial wireless communication, electronic warfare systems, semiconductors, and cable communications. The company was founded by Professor J. David Rhodes, a professor of electrical engineering at Leeds University. Education must be perceived as a tool that can define and establish the relationship between what is taught in school and its application to industry and the world of work. Such programmes can serve as a vehicle to develop linkages with industry through mentorship and coaching programmes. These programmes can encourage professionals to interact directly with educational institutions by offering part-time services through lectureship at secondary and tertiary levels to students who can benefit from their knowledge, experience and guidance in their career choices.

Institutions should ensure that every child is exposed to the teaching of some aspects of science and technology regardless of occupational choice so that there will be an appreciation of the application and role of science and technology to life. Governments should also develop or strengthen Technical and Vocational Education and Training (TVET) programmes and other technical programmes (private or state-operated) to encourage the development of the required skillset. Certification in the appropriate competency-based level skillset must be germane to the needs of industry and society. In other words, the need for technicians, machine operators, auto mechanics, maintenance engineers and related skills can be provided through well-structured vocational training and not necessarily through a graduate university programme. A diversity of skillsets will enable small states to diversify their resources into new and emerging sectors and to provide the basic skills that are needed for development.

There should be established relationships with the private sector so that they can participate in national manpower planning exercises and assist in the development of appropriate policies in education and R&D. This approach will ensure that manpower needs are identified, rationalised, quantified and coordinated. To give effect to this approach, the government should review existing legislation to allow for the movement of skilled labour within a specific regional grouping so that it would not put any country at a disadvantage when competing for foreign investment. The state should also mandate foreign companies operating in the local market to consider hiring a percentage of local staff with the necessary qualification and competence to their management team to enable local personnel can get the experience and develop the relevant competence in specific areas which could benefit future development.

Technology Transfer Policy Imperatives

The transfer of technology can be applied in a wider context to involve all aspects of the transfer of scientific or technological knowledge from one society to another to satisfy the identifiable needs of customers. Few countries are self-reliant on the technology they need; therefore, global transfer of technology has become an inevitable phenomenon across frontiers.

The design of machinery and equipment, for example, may have to be scaled and retrofitted to local conditions and indigenous resources may have to be adapted to the imported know-how. The 2008 study on Competitiveness and Innovation conducted by UNIDO and the Centre for the Development of Enterprise (CDE) pointed out that there are inherent weaknesses among Caribbean states regarding the availability of scientific skills, procurement of technology, and the capacity to innovate. Many of the problems in these areas seem to emanate from the knowledge and education systems.

The prevailing circumstance in many developing states regarding the modernisation of production plants and equipment suggests that small enterprises do not have the resources to upgrade their operations to make them technologically relevant to their needs and to enable them to be more competitive. Historically, evidence has shown that in some developing countries, a tremendous amount of resources had been expended on technologies that were unsuitable to the needs of the enterprises and the country. In many cases, there has been no plan to minimize or ameliorate the negative impact of technology transfer concerning the redundancy of workers, environmental impact, the economic application of the technology, training needs and maintenance. When the foregoing occurs, small developing states would become the victims of technology (embodied in equipment, machinery, and the like) that was not envisaged and planned for.

Furthermore, technology transfer mechanisms should be established to guide procedural matters towards the delivery of services (technical expertise and training), provision of relevant information, procurement of equipment and contingency measures to enable firms to maximize the benefits to be derived from given technologies.

When technology is transferred and made appropriate to the local conditions, it should be applied in favourable business environments to add value to indigenous resources that will transform them into finished products and thereby enhance the competitiveness of the relevant sector. It is not reasonable to think that technology should be made appropriate when it has already arrived in the importing country. Prior to the importation of technology, there should be a process of vetting and approval of the technology by skilled professionals supported by relevant documentation on the transfer details. This process should involve requisite planning, costing, social, cultural and economic impact assessment, including end of useful life disposal of the composite parts so that the appropriate mitigation measures and social safeguards can be initiated.

New technology has the possibility of creating subsidiary industries that can thrive on effective marketing approaches and techniques, skills development and international networking. The introduction of new technologies should be in keeping with the needs of an organization. Hence, the productive propensity of the technology should be given consideration when deciding on its implementation. This approach will avoid the underutilization of capacity, which will result in financial resources being expended in unprofitable ventures. Similarly, downstream activities should also be anticipated and integrated into a wider strategic plan, and the appropriate policy provisions should be made for displaced workers and their families as a result.

The needs of an organization relating to the productive propensity of the technology should be given consideration when deciding on its implementation so that there would be no underutilized capacity that will result in financial resources expended in unprofitable ventures. Similarly, downstream activities should also be anticipated and integrated into a wider strategic plan, and the appropriate policy provisions should be made for displaced workers and their families as a result.

To address the foregoing concerns, policy initiatives are required to ensure that an importing country can benefit optimally from the process of transferred technology without falling victim to any injudicious transfer mechanism. It was mentioned earlier that a useful approach to transferring technology is through Foreign Direct Investment (FDI). Through this medium, the foreign partner, who sources the technology assumes the risk upfront. It is imperative that governments establish relevant legislation adumbrating operating policies and conditions to facilitate the successful transfer and diffusion of technology. This will ensure that firms are not disadvantaged by being charged excessively high prices for technology that are obsolete or no longer state of the art, but which are ideally suitable for the needs identified.

There is a long history of derelict machinery and equipment that have populated the roadsides and open areas in many developing states because of a lack of spare parts or technical expertise to perform maintenance work on them. Such unsightly scenes have become the scourge and eyesore within the landscape of many developing states.

It is the responsibility of the government to devise policies that would ensure that any technology that comes to its shores does not result in derelict relics that are environmentally unsightly and damaging to the country; there may be trapped toxic gas which is part of the operating system for those equipment and machinery, which may require specialised equipment to remove them. Disposal of end-of-life machinery has also become a disposal nightmare for many small states.

R&D a Catalyst for Product Development

As was pointed out earlier, it will be imperative to streamline the educational system to incorporate more research and development programmes at critical levels of the system. Such an approach should have the support of governments, private sector organizations and research institutions. Research and Development should be encouraged to establish the environment for attracting and nurturing new talent and thereby ushering in the next generation of innovators which would provide the basis for investments to mushroom.

Several universities, colleges and research institutions serving selected regions in small developing states continue to defend their position as teaching institutions with limited research capability. On the other hand, some dedicated research institutions are woefully understaffed, under financed and lack clear directives and linkages to undertake R&D and promote technological development geared towards industrial development. Institutions should be adequately funded and supported to develop the relevant capacity and linkages to specific client needs. R&D is an expensive undertaking and realistically small states cannot afford to fund extensive R&D programmes.

For the most part, the research effort is essentially uncoordinated and there is no mechanism in place to disseminate research findings throughout the industrial sector. Institutions must therefore concentrate on research niches where they can develop comparative advantage so that greater use can be made of indigenous resources available.

In building a dynamic industrial sector in which R&D will play a pivotal role, developing states should be encouraged to establish technology extension services to provide linkages between research institutions and the private sector. Extension services will allow information from the result of research to be disseminated to the end-user of technology so that it can be used for value-added products and services. Enterprises that have the propensity to apply the result from R&D for product enhancement or new product development should establish the necessary manufacturing systems and processes as well as the appropriate skills to take advantage of such information. The necessary incentives should be provided for large private sector firms through specially designed incentive schemes that will provide funding for targeted research leading to value-added or product innovation in industry.

Intellectual Property (IP)

Intellectual property (IP) refers to creations of the mind and can include copyrights, trademarks, patents, industrial design rights and trade secrets in some jurisdictions. It has become one of the most important pillars of competitiveness in enterprises in a modern economy, particularly for highly innovative companies.

Intellectual Property allows people to own the work they create distinctively embodied in the form of a brand, an image, an invention, a design, music, art or another intellectual creation. It involves a system that provides legal protection of intangible assets arising from the creativity and innovativeness of individuals that constitute their rights. Several cases have been highlighted in the Caribbean and elsewhere where developers and inventors have lost the right to retain ownership of their inventions because there was a gross failure or neglect to protect the work that emanated from their creative ideas.

There are a select number of cases in which there was a failure to protect the brand in question and where foreign entities have patented those brands. The development of novel projects from indigenous materials to solve technological problems in some small developing states is still unfulfilled. This is evidenced by the dearth of patents from the region during the 20th and 21st centuries, a period in world history when technological development is at its zenith. It is reported that over 90% of the patents granted by developing countries is foreign owned. Very few of these patents are employed in the countries that have granted them.

In 2002, the alarming news broke that three patents were granted by the United States Patent and Trademark Office to two Americans namely George Whitmyre and Harvey J. Price for their "Production of a Caribbean Steelpan." The patent to produce the steel pan was based on a mechanised manufacturing process using hydroforming (a metal fabricating and forming process commonly used in car manufacturing to shape metals), as a cost-effective way of using high-pressure hydraulic fluid to mould malleable metal into lightweight, strong and complex shapes. This technology was different from the hand-crafting method which had given birth to Trinidad and Tobago's steelpan, the first musical instrument of its kind in the world.

The reason cited for failure to patent the invention was the lack of resources and the complexity of the process to do so. This situation therefore allowed foreign entities to capitalise on the opportunity to patent the invention and secure the rights and benefits that would be derived from this most creative invention. Since no one holds a patent for the steelpan that originated in Trinidad and Tobago in the early 1940s, the government appealed the decision and was granted a patent for the G pan in 2013.

Similarly, many years ago, the government of Guyana accused owners of Bedessee Imports Inc. a company founded by Guyanese living in Canada of unlawfully using the name "Demerara" and a map of Guyana on sugar products that were not made from Guyanese sugar. Among the products marketed by Bedessee include raw cane sugar from Mauritius, which it sells under the trademark 'Demerara Gold'. The label also contains a map of Guyana. Bedessee also markets a brown sugar product described as 'Guyanese Pride,' which also has a map of Guyana on the label. Neither product contains Guyanese Demerara sugar. This company had been using "Guyana Gold" as a trademark for its sugar and other products sold in Canada and the USA since 1984.

On the other hand, GuySuCo, the Guyanese company which manufactures the sugar in Guyana sold the product in 50 kg bags in the Caribbean market without branding the product. In 2003, the company launched its first brand for the retailed market under the brand name "Demerara Gold." The identical name that was used by Bedessee. After several legal challenges, it was decided that Bedessee, the first company to use the trademark will continue to use the mark on cane sugar products within Canada, Mexico, and the United States of America, whereas GuySuCo will continue to use the mark elsewhere in the world.

These cases occurred because the inventions or creations were subsequently registered by another party who claimed ownership of the inventions. This is an age-old problem that needs to be addressed as a matter of urgency, by providing the relevant information to would-be inventors or developers as well as creating and empowering the appropriate institutions to perform the necessary functions. In so doing, the industrial sector within a given region can experience some level of resurgence in the form of new and creative products. Innovators need to retain ownership of their creative work so that they can proceed to develop successful business enterprises.

Ownership of intellectual property can be facilitated by the establishment of a Patent and Trademark Office to serve a designated region. This will negate the need to register patents and trademarks in individual states. Apart from providing regional coverage, this proposed office will provide the necessary technical assistance for patent registration, which will result in reduced cost, improved efficiency and an established regional patent database. Furthermore, small states can provide the appropriate framework and instruments to stimulate and support innovation and the development of technology by building an indigenous industrial sector through the protection of the Region's intellectual assets.

IP can also be used as a vehicle to promote and maintain the competitiveness and authenticity of products and services particularly from regions such as those in the Caribbean, Pacific, or African regions to ensure that a small state's indigenous inventions are not re-branded by foreign entities and that it can maximize the benefits of its creative endeavours. This can only be possible with increased awareness about IP and its role and benefits. The nexus between IP and industrial development should be established, so that IP can be used as a mechanism to encourage and protect talents and skills as well as the indigenous resources.

Marie Sharp is an agri-based product manufacturer from Belize. She developed a successful pepper brand and traded under the name

Melinda's Hot Sauce[2]. She has been in the business since 1981. Over many years, she developed the business by putting in the necessary infrastructure to increase production, thereby enabling her to satisfy the growing demand for the product. She initially sold her product through a Belizean American importer in the United States. She however, neglected to get the American trademark on the name of the product and thought that he (the importer) had done so on her behalf, but instead, he had trademarked the product under his name. After spending several years developing the business, her marketer basically stole it from her by taking the trademark. Once the brand was well established and there was a thriving market for the product, the distributor who was marketing the sauce got the recipe, found someone in Costa Rica to make the product and sold it cheaper than the initial product. As a result, the Belizean entrepreneur was eliminated from business.

After a five-year struggle to reclaim ownership of her product, Sharp gave up the name Melinda's in exchange for being released from her exclusive contract with the distributor and re-branded the product under her name. The brand was the valuable factor in that case, and it must be noted that branding is everything in the marketplace today. Branding and intellectual property law are the key to product ownership. The Belizean woman's bioprospecting tale had a happy ending. It had taken her 20 years to rebuild the business to the success it is today.

The use of IP can discourage competitors from illegally copying products or services of other inventors and thereby benefitting from their work. As is often the case, products and services are replicated by other competitors without due regard to the resources and time expended by innovators in developing those products or services. Such practices undermine the creative capacity of the inventor as well as the ability to maximise the financial returns from the product or service. IP provides the requisite protection that would arrest this kind of behaviour. It can also enhance the commercial value of small firms, particularly in a situation when they are leveraging funds for further development.

---

[2] www.dallasnews.com/business

Investors can also be attracted to those enterprises whose products and services are legally protected. Moreover, financial institutions can become more amenable to providing loans to such companies.

Private enterprises can use patents, trademarks and copyrighted works as economic assets to generate income. Income generation can also be derived from technology licensing, franchising, trademark or copyright license agreements of the protected products or services. In the event of mergers or acquisitions, companies with IP can enjoy an advantage. IP can be ideally used as a marketing tool as in the case of product differentiation. As was discussed earlier, product differentiation allows a company to add value to its products or services. There are tremendous benefits to be derived from brand loyalty, which can be made possible through the use of IP.

In some regions, particularly those endowed with tropical rain forest and where a variety of tropical crops are grown, the local population should be mindful of the information they provide to foreign researchers who are engaged in bioprospecting which involves a host of unethical issues. The concept of "Bioprospecting" involves the discovery of biological resources from which new products are generated geared towards commercial purposes. Biological resources can be applied to the fields of pharmaceuticals, agriculture, and nanotechnology, among others. It is a general practice that large companies around the globe are scouring small countries for exotic ingredients and purchasing the rights to own them, thereby preventing the natives who have been using the ingredients for generations from earning a living from these exotic resources.

Business Associations

The importance of business associations (BAs) cannot be overemphasized in the case of small enterprise development. It has been recognised that small enterprises are more vulnerable than larger private sector entities. Co-operation among small business entities is critical to ensuring that they are represented at the policy level where their interests can best be addressed. At the governmental level, there needs to be someone to champion the cause of SMEs in a manner that will provide the kind of representation that will ascribe recognition and importance to the role they play in society and their contribution to the economy. BAs can provide direct services to the businesses they represent, which those small entities could not otherwise afford individually.

The advantages to be derived from cooperation through business association can be substantially beneficial to small establishments and far exceed what individual groups may receive from pursuing their own interests. Business associations are ideal vehicles to forge cooperation among small entities in several ways, particularly through networking and advocacy. Through BAs, small enterprises can build the relevant capacity through training as a means of developing their entrepreneurial and technical skills. They can also build networking capacity to interface with regional and international development partners. Business associations can facilitate the exchange of information and technology, promote economies of scale through joint procurement activities and strengthen their position to leverage the necessary resources to sustain their operations. Against this background, BAs can be ideally placed to transform the SME sector into vibrant economic entities, which can play a more meaningful role in the social and economic development of small States.

A business association is also a means of developing strategic alliances with research institutions through which they can participate in the dissemination of information relating to new technologies, technology transfers and product development possibilities that they would otherwise be unable to access. To ensure the continued growth and survival of small businesses, BAs can facilitate the promotion of clusters among similar private sector groups. This approach will seek to build cooperation and collaboration among businesses on a cost-sharing basis. It will also allow them to reduce costs and maximise their resources. BAs can promote a shared information technology database through which information pertaining to market studies, technology transfer for specific applications, technical assistance availability and the source and availability of financing could be accessed by participating enterprises.

Small enterprises should be encouraged and assisted in developing vibrant BAs among the various sub-sectors that will promote ways to create synergies, as well as to represent the various interest groups at the national and regional level on matters that will promote their development.

Infrastructure Development; A Key Prerequisite

For many years, small enterprises throughout many developing states around the world have been seriously constrained by the lack of adequate infrastructure including roads, power, water supply and other facilities that are generally considered basic to industrial development. More often than not, where these basic infrastructures are present, they are generally unreliable, the quality may be substandard, relative to what is required to meet modern standards and the cost may be prohibitive. Infrastructure such as roads, water and electricity often require upgrade and proper maintenance to provide optimum benefit to users.

For newer industries that are technologically driven, there is a need for infrastructure of the non-traditional type, particularly in the area of information and communication technology (ICT). Information communication technology is one of the key drivers to propel industry forward. To facilitate the provision of a modern telecommunication sector, service providers must make available to the business sector the requisite bandwidth, and other relevant technological infrastructure.

These infrastructure provisions will allow telecommunication service providers to deliver adequate, affordable, and reliable service to their customers. In some states, infrastructure programmes are being developed and implemented to address these shortcomings. For example, underwater fibre optic cables had been laid among some of the small island states of the eastern Caribbean. This initiative will provide the service providers with the required bandwidth to enable them to deliver faster and more efficient telecommunication services at affordable costs.

It is financially onerous for small enterprises in poor developing island states to raise the resources to invest in equipment and infrastructure that will enable them to achieve sustainable growth and become competitive. For the most part, governments are unable to adequately finance infrastructure costs such as water, energy, and telecommunications because many of them are highly indebted and lack the capacity to undertake further debt commitments. On the contrary, the cost of not making the requisite investment can be exorbitant from the point of view of opportunity lost in the absence of these fundamental services. For governments to cover these investment costs, they will have to impose a user charge on those enterprises using the services.

Whereas industrial policies are important to advance the prospect of industrial development, appropriate institutional capacity and regulations are also required to take advantage of opportunities to exploit international trade opportunities. For example, over the past three decades, Mauritius has effectively used industrial policy and trade preferences to facilitate the development of its manufacturing industries and to increase exports. It is instructive to note that there is overwhelming evidence to support the view that investment in critical infrastructure has become the catalyst for the growth of key industries.

The efficiency and transparency in customs operations, as well as the establishment of the relevant infrastructure to facilitate the transit of goods are critical for the enhancement of trade. This is one of several interventions that is necessary to promote a modern and vibrant manufacturing sector. At the macro level, the government should incentivise the development of industrial parks facilities for cluster development if deemed necessary, and logistics infrastructure like air-cargo complex, inland container depots, warehousing and logistics hub. This should be done in collaboration with the private sector.

The inadequacy of infrastructure is a serious impediment to national development in many developing states. Indeed, limited access to consistent and reliable power, water, telecommunications, waste treatment, and other public utilities and services, are major impediments to investments in the manufacturing sector. Although there has been a marked improvement in some small island states over the last recent decades, there is still a lot to be done to improve the relevant infrastructure and thereby create a favourable environment to facilitate the growth of industry. Small developing states need a sound and robust industrial policy with provisions to establish a broad based infrastructure that will transform the traditional industrial landscape such that there can be a diversity in the methods of producing goods and services that will redound to social and economic transformation. An industrial policy is therefore a sine qua non for national development, which will result in discernable economic and societal changes including urbanization and rapid employment. Investments that are novel and emanate from sound policy initiatives could be costly, but the return on such investments could have exponential benefits.

Financing and Incentives Requirements

Traditionally, most commercial banks in developing states have been established to provide financial and other support for the development needs of the local economies in several development areas. Over the years, these economies have expanded and have also diversified into several non-traditional activities including energy, information and communication technology and other services including culture, film, and music. Banks and other financial institutions have been reluctant to provide financing to these non-traditional sectors primarily because of the high risk associated with some of these new ventures. The lack of information about possibilities associated with these sectors as well as the lack of the in-house capacity to assess the viability of these projects and to quantify the credit risk associated with them cast doubt on the bank's willingness to provide financing. Consequently, several good development projects with commercial potential have been neglected.

In the past, many industrial projects have been funded by foreign investors or donor funding. Some countries devised new and innovative ways of providing funding for the manufacturing sector, particularly the SME sub-sector in the form of credit guarantees, microfinancing and other financing schemes through institutions such as development banks, credit unions and other credit facilities.

Notwithstanding the foregoing, there are other constraints that must be addressed to enable entities to acquire the resources necessary for development. In the case of regional and sub-regional groupings, harmonized investment policies can incentivise and stimulate private sector development through foreign investment flows that can finance private sector projects. Investors are generally attracted to a set of harmonized incentive conditions governed by uniformed and coherent investment policies, laws, and regulations toward private sector growth, rather than having to face different treatment for different individuals or groups.

To provide financing for the needs of small enterprises, financing mechanisms should be designed based on the achievement of measurable benchmarks by recipient firms. Such benchmarks should include performance monitoring, enforcement and compliance criteria, regular reporting, periodic review and impact assessment. To make industrial financing accessible to all firms, there should be stipulated criteria for the lending programme that must be executed transparently. At the regional level, governments within a given regional grouping should establish an integrated credit bureau to provide information to financial institutions concerning the creditworthiness of individuals and businesses. The loan portfolio of foreign firms should be monitored and controlled so that the liquidity of commercial banks would not be reduced to critical levels because of the acquisition of investment loans.

It will be useful to develop and implement harmonized investment policy framework (as against a fiscal incentive regime that exists in several states) to apply those elements that would contribute to a competitive investment climate. Macro-economic policies should be reviewed to create the framework for investment by encouraging business entities to have access to domestic credit at affordable interest rates.

Policy Alternative to Incentives

For many decades, some governments have relied on the provision of incentives to attract foreign direct investment in the manufacturing and tourism sectors. After many years, much has not changed. Some countries continue to offer incentives including tax holidays and duty-free imports on materials and equipment used in the refurbishment or construction of hotels, apartments, or any other business-related physical infrastructure. They also allow profit repatriation to attract foreign investors to their shores. Some multinational investors have even been granted preferential treatment of varying types. For the most part, incentives are discretionary and provide no guarantee for would-be investors. Studies and experience have shown that these incentives are not economically sustainable, and many foreign investors abuse these gestures of goodwill. A good functioning macroeconomic environment is more effective than investment incentives. Other measures have proven more cost-effective than granting those generous incentives of the types often provided.

Contrary to the view that the main stimulus for investment in industrial development should be linked to an attractive incentive scheme, it must be noted that an effective incentive scheme should be designed within the framework of a favourable and enabling environment with all the available elements of support to develop and promote the infusion of new ideas and innovation. The importance of the role of the private sector should also be recognised and encouraged to achieve growth within the economy of each state. Governments should also understand and fulfill their own role in creating the environment within which that growth should take place.

Research conducted by the World Bank Group's Investment Climate Advisory Services on the efficacy of investment incentives concluded that the incentives mentioned earlier are ineffective, especially where investment climates are weak, and they cannot compensate for such deficiencies. Even where incentives are effective in attracting investment, they have significant costs associated with them. When incentives are ineffective, political considerations often drive their continued use. Lower taxes should not be used as a means of compensating for a weak investment climate. Doing so results in revenue losses because incentives do not necessarily foster additional investments in such an environment, so the benefits go to investments that would have been made anyway.

The investment climate should be such that it influences the effectiveness of fiscal incentives in attracting investment through the role those public goods play in improving investment returns. In this case, public goods such as infrastructure, rule of law, enforcement of contracts, and so on, are components of the investment climate. The public goods are funded through a tax on capital, which in turn reduces the return on capital. But if public goods make capital more productive, then an increase in taxation spent on them would have the opposite effect. Improving the investment climate is a better strategy for attracting investments. In other words, governments should pay attention to the policy issues discussed earlier.

Production and manufacturing standards should also be accompanied by regulatory reforms to address the cost of compliance regulations for small manufacturers. Regulatory reform and the associated cost savings can be more important to some manufacturers than the incentives provided by various governments. Surprisingly, some of the successful niche manufacturing firms receive little or no state support. The beneficiaries of these incentives tend to operate in the area of export incentive schemes, lower corporate tax rates, training support, construction costs, financing trips abroad and several other promotional measures.

There is no doubt that grant schemes of one form or another exist in many small states to assist small manufacturing enterprises. If for instance, a thorough evaluation of the impact of these schemes reveals unsatisfactory results, it may be time for them to be reviewed to establish clear deliverables and outcomes that the government can measure and evaluate. This would ensure that such initiatives are not used to support unproductive enterprises or are used for non-business related activities. Such an approach would also serve to address the perception that government grant schemes tend to favour large businesses.

One of the key roles of government should be that of an "enabler" in creating the environment to facilitate small enterprises to develop and strive. Ensuring that all the necessary infrastructure are established with the involvement of the private sector playing a critical role and the conditions for access to financing and other assistance are within the scope of these enterprises. As a short term policy, incentives can be best applied when such investments are geared towards exploiting the abundant natural resources of the country.

Fiscal subsidies aimed at stimulating economic growth in the form of tax concessions to key economic sectors such as tourism and manufacturing can further erode an already fragile tax base and increase the economic vulnerability of small states. These concessions are not the only way to attract much-needed foreign investment toward private sector development that will increase employment opportunities. In some cases, small states cannot justify the opportunity cost to attract these investments. However, as a long term policy measure, governments can lower barriers to encourage investments in the exploitation of natural resources.

Sea and Air Transportation

The geographic orientation of some regional groupings of small developing states like those of the CARICOM Community makes transportation a critical factor in the movement of people, goods and capital. In some regions, the transportation infrastructure is underdeveloped, inadequate and ineffective.

The cost of transporting raw materials and finished products by air is usually prohibitive, particularly in a region where there are few air carriers. Alternatively, a well organised and efficient sea and air transport service may improve inter-regional trade and thereby contribute to the improvement of competitiveness among industrial enterprises. Currently, the transportation system in some regions designed to facilitate intra-regional trade can be generally described as unreliable, ineffective and expensive.

Transportation is a major requirement for the movement of products to various market destinations around the world. The demand for air freight is limited by cost. Air transport is about 4 or 5 times that of road transport and 15 times that of sea transport. Commodities shipped by air generally have high values per unit cost or are very time-sensitive, such as perishable agricultural produce and seafood.

However, in the context of business, transport must be efficient and cost-effective to link the supply chain with markets. Transport services as a regional public good include subsidiary infrastructure such as specially designed transport with cold storage facilities and holding and shipping facilities. An efficient intra-regional transport service is a necessity to transport the volume of goods among the various island states. It would be difficult to justify such investment because the volume of goods to be transported may not make the investment financially viable. Other issues to be resolved include the choice relating to the modes or mix of the transport system, the role and participation of the private and public sectors, whether goods only, or goods and passengers and the question of financing deficit during the development stages.

One of the factors that significantly impact freight charges in most regions is the cost of outbound containers. In some regions, because of the decline in agricultural activities and limited scope in manufacturing production, the volume of bulk cargo for export trade is low. As a result, a high proportion of the containers leave the ports empty, which effectively has a cost associated with zero volume. The cost of transportation has a significant impact on key sectors like tourism, agriculture, entertainment and manufacturing.

There are other factors including labour and logistical issues that inhibit cost reduction technologies and strategies within the maritime system which can otherwise engender lower cost and effective port management. The constraints within the sector should be addressed by taking several strategic actions as a matter of priority and with urgency.

Small states should encourage investments particularly by the private sector and possibly through cooperation initiatives in logistics infrastructure. All maritime operations should be examined to identify possible bottlenecks and inefficiencies and then adopt best practices in port operations to improve efficiency, minimize costs and thereby facilitate trade among states. Adequate access to ports through the provision of a greater flow of cargo utilizing the most cost-effective logistical arrangement should be provided.

A study conducted by an Asian Development Bank revealed that among the constraints that retard the growth of niche markets in most Pacific Island Countries (PIC) for high-end products like pearls, vanilla and cocoa is the cost and inefficiency of sea and air transportation. Since exorbitant transportation costs negatively impact export competitiveness by increasing the time and costs involved in manufacturing these goods, the PIC governments should therefore pursue substantial investments to improve their transport infrastructure.

The major advantage of air freight is shortened transit times. The time taken for cargo operations could complement expeditious transit time or undermine it. The time taken for goods to become available for trade will depend on several factors including customs clearance procedures, cargo inspection procedures, the efficiency of cargo handlers and the layout of storage facilities. The foregoing factors as well as custom charges on imported raw materials should be reviewed so that the cost of production would not be compromised by these upfront charges. It is necessary that all existing labour laws that affect modern port operations be reviewed and revised where necessary in keeping with modern port and customs operations.

Public-Private Sector Partnership

Public-Private Partnership (PPP) can generally connote a broad range of agreements that set out the terms and conditions of cooperation between public institutions and the private sector. Such arrangements seek to build on the assets of each partner that best meets defined public sector needs through the appropriate allocation of resources, risks and rewards. PPPs have often resulted in the transfer of investment projects to the private sector that have traditionally been executed and financed by the public sector. This process entails bringing together social priorities with the managerial skills of the private sector, relieving the government of the burden of large capital expenditure and transferring the risk of cost overruns to the private sector. Rather than completely transferring public assets to the private sector, as with privatization, the government and the private sector work together to provide public goods and services.

Because of the nature of both the public and private sectors and the objectives of the constituents they serve, both sectors are replete with challenges and incompatibilities that are often construed as weaknesses. The public sector operations are conceived to be less efficient than those of the private sector. It is also reputed for discriminatory behaviour, tedious bureaucratic red tape, insufficient transparency and difficulties in termination of contracts for underperformance, as well as failure to deliver on promises.

The private sector is also reputed to strive on its monopolistic status that propels exploitation by price hikes, supposedly justified by the requirements for profits. It is also renowned for job losses and insecurity following involvement in the pursuit of productivity improvements and cost reductions. The private sector exploits situations where the necessary regulations are weak, not in place, lack transparency or are vulnerable to political interference. For example, they resort to private incentives to bypass environmental conservation requirements or ignore other public interest concerns. It is for these foregoing factors that both sectors when they combine their resources, can achieve tremendous accomplishments.

As is often the case in developing countries, a Public Private Partnership is sometimes a contractual agreement involving a mix of a public agency (state, local) of the host country, a private sector entity or consortium of firms or private foreign or domestic investors, a foreign public agency such as the Foreign, Commonwealth & Development Office (formally DFID), European Commission, and USAID and an international organization such as the World Bank, or other multilateral agencies including a United Nations Agency, or a combination. Through agreements forged between governments and these agencies, the skills and assets of each partner are shared in developing and delivering a service or facility for the use of the public. The forms of partnerships may include joint ventures, PPPs, privatization, outsourcing, or a variety of collaborative or cooperative arrangements.

Political expectations may tend to influence the range of activities in which PPPs are engaged. Some countries have expanded the use of PPPs from infrastructure projects like roads and bridges, power generation and distribution and telecommunications to traditional core public services such as healthcare and hospitals, housing, postal services, ports, waste and water management. In other countries, these core public services are either still reserved exclusively for the public sector or are developed separately by both public and private sectors.

For many development projects, the private sector involvement, which is driven by the profit motive, can help to ameliorate the constraint of insufficient public sector capital and expertise for costly physical infrastructure projects that are by nature considered the basic long-term social responsibility of government. In most developing countries, the impact of large budget deficits, heavy debt servicing obligations, limited borrowing and taxable capacity, huge and increasing infrastructure deficits have greatly enhanced the burden of the state and have limited governments' fiscal space. The PPP model is therefore conceived as an attractive option for relieving the government of the massive financial burden of defraying the cost of those responsibilities. The private sector is known to possess the requisite resources, management expertise and technological innovations capacity to absorb operating and maintenance costs that would otherwise become the responsibility of the government.

Private partners can bring greater market efficiency in the form of competitive bidding and other procurement methodology for contracts and by introducing several cost-cutting measures to limit waste which constrain governments due to political interference. The traditionally accepted views that certain public goods and services such as education, including vocational training, water and waste management are to be freely provided by the government, should be rationalised with the accepted fact that many of these services are taken for granted and are thereby abused. The involvement of the private partner will ensure that some aspects of these services are paid for and that there is accountability and cost sharing for these services. In general, the operation of competitive market forces to which private enterprise is subjected is thought to bring to the partnership greater efficiency and expertise. There is a body of evidence which suggests that PPPs have accomplished remarkable gains in efficiency, cost reduction and expansion of service over time.

PPPs can also play an important role in reducing liability risk. For example, a public utility wholly owned by the public sector but managed by private sector operators, who have no equity interest in the utility but would have an interest in operating the facility more efficiently. Using a PPP model would spread the liability risk among the partners, and thereby reduce the exposure to such risk for both partners in the venture.

It can be argued that the PPPs have a better record of environmental protection and enhancement than enterprises operated purely as private ventures. The reason for this is that the private partner in a PPP then has a stronger motivation to comply with government regulatory standards, as it is in the interest of the private partner to maintain a good reputation for ensuring compliance using better technology and effective management. Evidence has shown that in developing countries, the use of PPPs often results in the expansion of the provision of public utility services, such as water supply and electricity connection, particularly to poor urban communities. This can be promoted by structuring the PPP contract in such a way as to ensure such an outcome is feasible. The private partner often finds that it is in its interest to exploit any networking connections that could make for the profitable expansion of services.

On the evidence of experience and capacity, PPPs can institute the relevant framework for effective human resource management and improvements. This is driven by the private partner's motivation to comply with government and international industrial health and safety standards, as well as its motivation to employ an effective management strategy to avoid worker grievances and resolve conflicts to mutual satisfaction, which can otherwise be costly if these matters are not given priority.

It is in the interest of the government to establish the framework for functional cooperation with the private sector in the formation of a public private partnership. Such partnership should be governed by a set of established agreements and conditionalities that will be attractive to private entities, bearing in mind, the profit motive of private sector entities. The need for cooperation is predicated on the notion that governments, which implicitly represent the public sector lack the requisite resources and technical skills to forge the development agenda ahead. This kind of cooperation must address the needs of the public sector for targeted development.

Traditionally, attempts at fostering PPP in the Caribbean and elsewhere have not had a high rate of success because of a lack of trust in most cases due to political and other reasons. This has invariably led to the failure of meaningful engagements by both sectors. For the most part, the forms of partnerships that have become the norm are not identified and as a result, there is no clear delineation of roles and responsibilities. The foregoing factors often result in the government having to develop projects for which resources and skills are not adequate or are not available.

Public Private Partnership will require practical mechanisms and modalities on which to establish cooperation. Such cooperation should be built on what is considered best practices. Collaboration between the government and the private sector can result in the development of basic infrastructure like roads, power, water supply, housing, industrial parks and other national projects. PPPs can also provide the framework for the establishment of a network of support among regional and extra regional institutions, entrepreneurs and business entities, through which there can be an exchange of information and sharing of experiences, industrial coaching and mentorship programmes. Through this cooperation, a forum could be established for periodic policy dialogue between the public and private sectors to address challenges as they arise.

The relationship between the private and the public sector should be based on trust, which will allow the private sector to participate meaningfully in key development projects. Through their involvement in these projects, they can assist the government to build capacity and to share expertise and technologies in critical areas such as management, finance, marketing and engineering. On the other hand, the government can assist small firms to develop relevant capacity in key areas to enable them to capitalize on outsourcing or subcontracting arrangements by larger firms both foreign and local, thereby creating their own niche. The private sector is renowned for skills that are not common in the public sector. Such skills can be used to provide support in project activities where there is a need for superior risk management and project management expertise to effectuate the implementation of projects.

Energy for Sustainable Development

It has been over five decades since the world experienced the first major oil crisis in October of 1973. The Middle East region still controls a little less than two-thirds of the world's oil supply. The demand for energy particularly from developing countries like China and India, as well as higher oil extraction costs continue to increase, although a few years ago prior to the onset of the Coronavirus (COVID-19) pandemic, there was a marked decrease in the commodity on the world market.

The role of energy and its impact on every aspect of development is well known and documented in most developing countries. For many centuries traditional energy sources have been exploited to fulfil mankind's basic energy needs. Today, most of these natural reserves are under threat of exhaustion and depletion. New and emerging energy technologies including fuel cells, solar PV technologies, electric mobility, solar thermal and hydrogen systems, among others, should be exploited responsibly to provide energy solutions to the growing demand of society.

Indeed, traditional fossil fuel-based sources that have been primarily used to provide energy services are known to have created serious environmental problems, which have seriously impacted many small developing states because of their physical vulnerability. Economists have struggled to understand the ways that disturbances to the supply and demand balance in energy markets influence economic growth and inflation. Some developing countries are therefore grappling with critical issues relating to the security of supplies and the cost of petroleum products traded in some regions.

Before the reduction in the price of petroleum on the world market in the mid-2000s, there were fears that the continued upward movement in the market price of petroleum and by extension, its by-products including gasoline, diesel, jet fuel and propane. An increase in the price of petroleum and its by- products on the world market would seriously impact various development sectors in small developing states including tourism, transportation, agriculture and manufacturing. There is a direct nexus between these energy inputs and the performance of major traditional sectors in most developing countries such that price variations in the market have caused serious difficulties in small island economies.

Higher energy prices have economic implications for every country. In the long run, higher energy prices are likely to significantly reduce the productive capacity of small island states' economies. It is evident that high energy costs have made some businesses less willing to invest in new capital to the extent that some existing capital have become obsolete. This would also impact the economy in ways that would restrain the growth of labour productivity, which in turn would reduce real wages and profits.

It cannot be overemphasized that sustainable energy holds the key to tackling the twin scourge of poverty and ecological degradation. Small states must therefore begin to focus concomitantly on how they use energy as a tool to promote sustainable development within their frontiers as well as to fulfil their global commitments. Small developing states must understand the nature and scope of their development within the global context and the nexus between social and economic development. Moreover, there must be an understanding of the complexities of the issues involved, as society transits from agrarian to more diverse economies as dictated by the new global economic order. Energy services are critical deliverables to society and not necessarily fuel or electricity.

More than half of the world's population lives in rural areas and does not have access to labour-saving and value-adding services that electricity enables. Society is interested in household lighting, cooking, refrigeration and entertainment for its advancement and survival. In any expanding and progressive economy, the need for energy per unit of GDP increases. Access to modern energy services can alleviate the drudgery in the lives of millions of people in developing states.

Higher energy prices are likely to reduce the productive capacity of the economies of small developing states in the long run. The impact of this would restrain the growth of labour productivity, which in turn will reduce the cost of real wages and there will be lower profit margins. The higher cost of imported oil is likely to adversely affect a country's terms of trade and its current account deficit. To provide for the energy needs of a vibrant industrial sector and to mitigate against continued rising prices in energy and any form of insecurity in supply, several actions may be required. The government should establish a regulatory body that will develop efficiency standards for equipment and electronic products that use electricity. No power company should be allowed to regulate itself. In this regard, the regulatory body should be independent of the state and other external influences.

A regulatory body should ensure that there is minimum loss of electricity from transmission lines as a result of archaic distribution and grid systems. Generation equipment that have come to the end of their useful life should be replaced with modern efficient equipment. As a matter of policy, energy-efficiency technologies should be encouraged by using the most current technology sources. New equipment, appliances and industrial plants should be guided by the principle of low energy consumption for cooling lighting, and heating.

Governments should enter into bilateral arrangements or multinational agreements with countries that can supply energy at affordable cost when it is mutually beneficial to them. This will ensure greater levels of energy security, so that small states can support development activities in the industrial sector sustainably and securely. To encourage competition in the market, some countries have regulated the energy market by removing or relaxing barriers that will prevent independent power producers (IPPs) from generating energy using various renewable energy forms. Furthermore, incumbent utility companies should be mandated (through well negotiated power purchase agreements) to purchase power from these providers when it is economical to do so. This could allow national grids to sell energy at reduced prices to the consumer as well as influence a reduction of the tariff on energy.

The government should provide incentives e.g., a "Green mortgage" for energy-efficient buildings, equipment, as well as for standalone scalable renewable energy systems such as solar PV, wind and hydro plants (mini, micro and pico) that will provide alternative energy sources for SMEs in rural communities. This will help to improve their productivity and encourage efficiency in their enterprise.

Financial incentives to lower the barriers of high up-front costs for specific energy-saving devices and renewable energy technologies should also be provided. Government should examine the possibility of raising resources to finance renewable energy projects by establishing energy efficiency and renewable energy fund to be financed by levies on fossil fuel imports.

All manufacturing enterprises should be encouraged to conduct energy audits to identify possible areas of inefficiency, which will result in energy loss to which there is an associated cost. Such losses will increase the production cost that is generally passed on to the consumers. Financial incentives could also be offered to enterprises that conduct energy audits.

Industrial Capacity Building

To survive in the current global marketplace, small enterprises should develop a new industrial paradigm that is driven by a philosophy of change. This approach should be responsive to the challenges of globalization and driven by innovativeness in technology, value-added initiatives, product design and marketing.

Over the years, there have been several initiatives aimed at stimulating growth in the manufacturing sector in some developing countries. Regrettably, many of these programmes have been repetitive and lack the diversity that encapsulates a new thrust in optimising anticipated results. New initiatives should therefore be devised against the backdrop of structural change and resource endowment. In so doing, enterprises would have a greater chance of competing successfully in the global market. At the macro level, governments must recognise the benefits of return on investment from a pilot project approach before they are fully developed into national projects. There should be a willingness to apply lessons learnt from successful development programmes in other small states of similar historical, political and social orientation. Small states must be amenable to undertake a transformation of the economy through structural change, diversification, and upgrading of the production systems.

There should be a demonstration of commitment to reducing the number of failed business enterprises on a regular basis. It is therefore necessary to have a framework of some kind within which to provide adequate support for them. Business Incubation has been touted as the remedial recourse for the failure of start-up businesses. A typical incubation system offers a variety of support services most of which are provided within a designated facility. Support programmes that are offered include training workshops in accounting, marketing, ICT services, counselling and advisory services.

Several reports have shown that most incubators in the Caribbean and elsewhere have experienced some level of failure. Whenever they have been deemed successful, they have become a burden to the state and private sector agencies. In most cases, incubators ceased to exist when development funds have dried up and further funding cannot be sourced locally. The time is perhaps right to rethink the philosophy of business incubation in the context of the small island developing states. Business support models can be developed in which the state, in collaboration with the private sector, can provide only those services that are critical to business start-ups offered by state agencies including Bureaux of Standards, Small Enterprise Development Units (SEDUs) and National Development Foundations. This approach could reduce waste that is often associated with the provision of free services, increase efficiency and help to develop an independent mindset with an entrepreneurial drive to stimulate SMEs to work towards acquiring assets that they can use as collateral in the future.

Incubators have had high success rates in several Asian and European countries, in that, they have demonstrated growth within enterprises. For smaller states, it is necessary for the concept to be customized to give rise to what could be called a "Business Support Programme". Within the framework of a Business Support Programme, several initiatives including incubation, clustering, and benchmarking where applicable and desirable, can emanate and enhance the possibility of success for small firms.

Where it is practicable, selected areas could be demarcated for the establishment of specialised industrial parks, export processing zones, and the like. In cases where they are considered to have growth potential and economic viability, the establishment of a business support programme should be encouraged to provide backstopping support in critical areas such as ICT training, standards and quality assessment and the development of product formulations. The needs of small enterprises should be identified by professional assessment so that the requisite support can be provided during the initial start-up stage of the enterprise. What is commonly called a "Cluster of firms," should be more appropriately referred to as networking among firms. This approach could be promoted where applicable, as an inclusive and effective strategy to build capacity among homogeneous enterprises through networking. This initiative will serve to identify challenges in SMEs, establish the requisite support infrastructure, enhance cooperation that will lead to sharing of information, technology and technical skills, thereby enabling them to take advantage of market opportunities.

Institutional and Legal Strengthening

Notwithstanding the middle-income status of many small developing states, and their relatively high per capita income, there is an inherent weakness in their institutional capacity when compared to other relatively small states such Iceland, Luxemburg as well as those in the western Balkans. Several reports emanating from studies on the industrial sector in the Caribbean for instance, emphasised the point that the lack of administrative and legal requirements have to some extent stymied the rate of progress of industrial development particularly at the initial stage. When combined with the bureaucracies that exist in most of these states, these requirements can present a daunting and frustrating experience for would-be entrepreneurs who want to "hit the ground running."

There have been situations where potential investors have been discouraged by challenges posed by the bureaucratic red tape involved in doing business in some jurisdictions. Some investors have been attracted to other destinations, where they have been accommodated because the administrative system is more amenable to facilitate them by expediting the administrative, legal, and other processes necessary for the establishment of their business ventures. Some small states do not have the legislative framework in place to deal with sophisticated technological infusion and are therefore unsure what requirements should be met regarding specific working environmental conditions, operational and safety guidelines, among others. In such a situation, governments have become less amenable to granting permission for the establishment of business enterprises, where the transfer of technology is a major activity.

Against the foregoing backdrop, the requisite legal and institutional framework will be necessary to address this matter. The process of applying for permission to conduct business in some small states is a long-drawn-out and bureaucratic one, which can frustrate would-be developers. A legislative strategy should be devised to expedite the application process for seeking license to conduct business in small states. In so doing, it will be expedient to initiate a system of "single point" contact or "electronic window," where all applications for business license are completed and processed electronically. This method would accelerate the process of approvals by providing the relevant statutory and administrative clearance within a short but predetermined time limit.

It may be necessary for states to undertake legal reforms to address issues that may be associated with new technologies that may likely impact the labour force and the environment and have possible social and economic consequences. This will avoid unnecessary delay and frustration in attaining the required approval. Relevant institutions need to be strengthened to provide the necessary capacity at the administrative and technical levels to adequately address the demands of the industrial sector, thereby creating an interface with other agencies that are critical to the provision of services.

Technologies of various kinds are becoming everyday tools in business, trade and finance. For example, trade in services will require the requisite legal platform to build confidence in the various processes. Addressing these challenges facing cross border e-commerce will in part require the government's intervention, preferably on a regional basis. Some governments have adopted measures to encourage firms to enter cross-border e-commerce and reduce user mistrust. Their export promotion agencies have carried out general training and trade facilitation programmes, developed logistics platforms and services, financed website design and marketing and introduced new cross-border payment methods. In addition, to foster trust in cross-border e-commerce, several national associations and chambers of e-commerce have promoted self-regulation through appropriate laws that promote codes of conduct.

Institutional strengthening involves, among other things, strengthening key institutions within the state including the justice, security and social systems. It requires enacting the requisite legislation to strengthen the functioning of the legal system to enforce the rule of law, thereby combating corruption, improving public administration, strengthening the regulatory system, and promoting effective social development policies. Institutional strengthening provides an opportunity to optimise performance in state institutions. There is a myriad of inherent weaknesses in the state institutions of some small developing states. In the absence of the resources and expertise to address these weaknesses, small states often depend on development assistance including technical assistance or funds that are often provided by developed countries or multinational agencies.

It has been widely reported that development assistance has not been effective in solving many of these institutional challenges, partly because many of them are rooted in weak organizational structures which engender poor institutional functioning. Moreover, institutions are generally renowned for poor management epitomized by a lack of accountability. The onus is on small states to organise their public institutions in a manner that will enable them to adopt international best practices in their operations so that they can respond to the needs of the people they serve.

Production Integration

Production Integration can be defined as the direct organization of production in more than one state by a single economic entity. Complementary production involves production among several economic enterprises operating in one or more states to use required inputs in the production chain. These activities lend themselves to cooperation among economic enterprises in areas such as purchasing, marketing and research and development. Production integration can also be considered as the cross-border dispersion of component production/assembly within vertically integrated production processes. In this regard, each country specializes in a particular stage of the production sequence, which has been an important feature of the deepening structural interdependence of the world economy in recent decades.

This collaboration is not restricted to manufacturing activities but also includes the services sector, capital, technology, and the like. Production integration which is characteristically similar to the global value chain in some limited context is intended to encourage entrepreneurs across a given region to actively consider joint venture investment in related industrial activities in other states, particularly where there are distinct advantages in various cost centres. As was pointed out earlier, the implicit and generalized industrial policy of some states can be veritably described as inward-looking, resource-based and state directed. This approach has failed with the new policy environment which has shifted towards economic liberalisation and outward-looking development strategies.

Notwithstanding popular rhetoric in support of production integration, there has been no serious or sustained effort by some regional governments to promote this concept. In the context of liberalised trade and globalism, traditional inward-looking economies should adopt more outward-looking development policies and strategies that can respond to the challenges and opportunities of the global market. Regional integration on which production integration is predicated offers several advantages concerning labour mobility, joint investment and promotion, more open trade and regulatory cooperation. For production integration to be successful, the necessary legal and institutional framework must be instituted to facilitate the establishment of the necessary platform that will foster prospective investor confidence.

The legal framework should be established to allow business enterprises across a particular region to pool their resources in the production of goods and services, utilizing inputs from countries within a particular grouping of states with a particular endowment. Enterprises can then maximise the benefits to be gained from economies of scale both in financial terms, access to technology, market expansion and employment opportunities.

Marketing and Distribution Policy

There is generally a disconnect in many industrial enterprises between what is being produced and what the market requires. It must be understood that an important aspect of competitiveness from the point of view of the marketing philosophy must involve the issues of being able to get the product to the market at the right time, to the right customers and at the right price, thereby expanding the market base. SMEs can hardly find the resources to do their own marketing on a wide scale or to engage the services of reputable marketing firms. Many of the regions in which small enterprises operate lack professional marketing and distribution agencies that can adequately support the manufacturing sector.

A marketing agency will allow products to be moved from the point of production to retail outlets and then to the consumer, thereby creating an interface with the consumer. SMEs need to have access to distribution agencies (with the appropriate channel of distribution) that will provide a range of services including marketing intelligence, addressing buyers' needs, product promotion as well as taking some commercial risks. SMEs may not necessarily need sophisticated channels of distribution because of their product range. Because of the limited size of the SME sector in many regions, the volume of goods they produce, and the costs associated with marketing and distribution, they must grapple with ways and modalities of selling their products. This exercise, which is generally time consuming, could distract them from the core functions of managing the plant operations and all the attendant processes. Alternatively, they have to seek the assistance of foreign marketing agencies at exorbitant costs. A marketing and distribution strategy should be implemented along with the appropriate policies designed to support small enterprises in the marketing of their products.

One of the strategies to address the marketing and distribution shortcomings in some regional jurisdictions is to establish formal cooperation with a regional entity such as the Caribbean Export or any other marketing agencies to initiate a dedicated marketing promotion that will coordinate the marketing and distribution of the region's products. This can be done by either working directly with distributors or in collaboration with other reputable extra-regional agencies to global destinations. Cooperation between the government and the private sector can result in joint venture initiatives to provide financial assistance to small enterprises that will enable them to participate in technology fairs, trade fairs, export promotion missions and exhibitions. Such cooperation should be structured in a way that will allow entities of similar product types to cooperate within an agreed framework and operational guidelines to satisfy production demand. SMEs should leverage the resources and build the relevant ICT capacity within the sector to equip them to market and transact business online.

Labour Relations and Good Governance

The importance of good industrial relations is often underestimated from the point of view of good governance. Good industrial relations ensure stability within the industrial environment and reflect good governance. Failure on the part of governments to establish the legal framework to address matters relating to working conditions, wages and benefits, the rights of the employer and those of the employees, could give rise to unbridled confrontation and unrest between workers (through their trade unions) and management, which could plunge a country into a state of chaos and anarchy.

Some governments have not demonstrated the willingness to curb the enticement of the drug trade, which is likely to compromise the effectiveness of state governance and render it weak. The social and economic cost of crime, and in particular, drug-related criminal activities can be catastrophic to small states at best and can be a disincentive to investment both foreign and domestic. The cost of poor governance can engender poor policy/decision-making, encourage corruption and breed administrative incompetence. These sociopolitical drawbacks will be devastating to small states or any country for that matter. Anecdotal evidence emanating from illicit activities in Columbia and Mexico are in themselves instructive. The governance of small states cannot afford to fall into such a state of decline because it will take longer to recover from such ignominious failure in governance.

Politicians are known to use their power to make decisions on matters that are beyond their remit to appease their political cronies in accomplishing their goals. Ideally, these circumstances perpetuate corruption in governance. In a small island state, a standard procurement process was severely compromised when there was political intervention to allow a foreign contracting firm to undermine a contract by reducing the scope of a project in lieu of providing financial assistance to buffer an increase in the market price for energy products. During the negotiation of a fixed term contract, the contracting firm recognised the increasing trend in the market price for specific energy products, occasioned by the disruption in the global supply chain due to the Covid 19 pandemic. The matter was raised, and the client generously acceded to the request by adjusting the contract price to take into account the price increase in the market. After the contract was signed, the contracting firm delayed the purchase of the equipment, which was contrary to what had been agreed upon. Subsequently, the contracting firm requested a second price increase, or alternatively, a reduction in the scope of the project.

The project management team vehemently rejected the request on the grounds that the contract had not made provision for any increase in prices, notwithstanding the client negotiated in good faith by granting the request for a price increase. Furthermore, the project management team argued that the firm had already received an increase in price during the negotiation process. Moreover, the project had limited funds available to implement the project. The contracting firm, through a political crony, sought the intervention of the political directorate. To appease that crony, the Project Manager was dismissed, the terms of the contract were abrogated, and therefore the scope of the project was reduced. The firm was allowed to build a smaller plant at a higher incremental market cost. This decision cost the country millions of dollars over the projected life of the project.

While new technologies represent an asset, they can also pose a significant threat if not used for the right purpose. Cybercrime is proliferating, and small states do not always have the adequate human and technical capacities to combat it. There needs to be a regional approach and action plan to tackle this most vexing issue including the enactment of relevant legislation.

Existing industrial relations policies should be constantly reviewed to make them modern in keeping with international norms and best practices. Indeed, such a system should also respond to new changes and developments in the workplace supported by new technology and innovative systems that will require adaptation and adjustment from all employees. Industrial relations should address matters relating to culture, history and the political realities of a country. It is also a function of trade unionism and must therefore address the issue of education and training, dispute settlement and increased productivity. It augurs well for sustained industrial growth and productivity in any region and is also an impetus for the promotion of a stable industrial climate, which is a prerequisite to encourage and attract foreign direct investment.

To develop an industrial relations system that will provide a sustainable industrial climate, enhance human resources and augment productivity, good governance should be promoted and practised at all levels of government. This can be done through ongoing engagement with civil society on a wide range of issues including the elimination of corruption in public places and the promotion of the rights and freedom of every citizen, among other things. The labour laws should be modernised to ensure that they are relevant to current trends in the labour market, particularly in the promotion of harmonious relationships in the workplace and demand and supply of labour so that the issue of productivity can be addressed in a manner that will allow industry to fully maximise labour inputs.

Traditionally, wage negotiation was the focus of collective bargaining. Indeed, collective bargaining should involve managing change in a socially acceptable way on matters of innovation policy, professional training, flexibility, the labour market and wage developments, employment policy and industrial policy aspects of restructuring negotiations. There should be an analysis of key trends underlying the development of the industrial sector, particularly concerning the dynamic structural change in the world economy, which will help to deepen the dialogue in managing change within the industrial sector.

Environmentally Sustainable Manufacturing

The best available scientific evidence confirms that there is enormous risk to the planet posed by the current production and consumption patterns in respect of human health and development opportunities for future generations. It is worth lamenting with regret, that the disappearing species of plants and animals; the destruction of forests, which has given rise to desertification; the proliferation of greenhouse gases in the atmosphere, which has resulted in global warming and the melting of the polar ice cap, which consequently leads to sea level rise, bear testimony of the effect of man's indiscriminate actions. The irreversible impact of climate change increases this risk to the point where there is great certainty that pollution and greenhouse gas emissions will reach a point of no return, thereby creating environmental imbalances.

Admittedly, the technological age has engendered a plethora of existing environmental problems that have been exacerbated beyond imagination. The lack of proper environmental management programmes and awareness has been identified as one of the critical factors that have obfuscated the consciousness of the public towards taking the necessary actions to arrest environmental degradation. It would seem obvious that the innovators of new technologies have focused on the supply end of the continuum, and in some cases, there are few built-in environmental safeguards for the end-users of technology. On the other hand, industry, which is a primary user of technology is not prepared to take responsibility for activities that are environmentally unfriendly and deleterious. In other words, risk assessment is not always factored into the technological equation as far as the interest of society is concerned.

Small developing states are among the group of countries least responsible for the increasing carbon emissions at the global level. Nevertheless, there seems to be a binding commitment from these states to reducing their CO_2 emissions, which is reflected in the overall improvement in their CO_2 emissions efficiency per unit of value-added. This decrease, along with relatively low investments in polluting technologies, and the rich natural resources of these small states (agriculture, forest resources, biodiversity, tourism, minerals and oil extracts) show that LDCs are well-positioned in the transition to a green economy.

Developed countries have generally incentivised small developing states to reduce their carbon footprint through carbon trading and other incentive schemes including the utilisation of non-renewable resources. This strategy allows developed countries that are major polluters to continue their wanton pollution of the earth and avoid their carbon reduction commitments in pursuit of providing the insatiable wants of their citizens. In so doing, little attention is paid to the plight of small island developing states that contribute a minimal amount of greenhouse gas emission, but in the long run, have become the victims of the deleterious effects of climate change.

To promote economic development, it is critical that SMEs commit themselves to participate in the transition to more sustainable patterns of production and consumption. In the context of small developing states, it can be argued that the individual environmental footprint of SMEs may be low, but their aggregate impacts can, in some sectors, exceed that of large companies. SMEs have a responsibility to contribute to the transformation of the environment into a green environment by reducing the environmental impact caused by their actions. This can only be done by ensuring that the activities in which they are involved do not contribute significantly to the pollution and the generation of waste that are the scourge of society.

The ever-increasing world population is making an unprecedented demand for material goods and services at a rate that cannot continue indefinitely without placing undue pressure on the finite resources of the earth. The technology-push concept of consumerism propels the emergence of new technologies, which creates a desideratum to provide alternatives through anthropogenic substitutes in quantities required by industry. It must be emphasized that most of the naturally occurring minerals are fast becoming exhaustive. Consequently, it is anticipated that a large proportion of the technologies designed to use alternative materials in the quantum required to sustain demand have negative repercussions for society and should be addressed urgently. For instance, the wanton recklessness in the disposal of wastes, pollutants and contaminants from the multiplicity of production processes must be of concern to humanity.

The comparative advantage of a country, especially during the early stages of its development, is usually based on its abundant resource or factor endowment. This explains the rate and manner in which factors of production are depleted. The use of selected technology cannot be blamed exclusively for the degradation of the environment. Many modern-day activities are driven by economic factors which also give effect to negative environmental practices. It could be argued that underdevelopment and poverty could be considered precursors to environmental degradation. These social ills have encouraged the destruction of large forests around the world, as the demand for wood for fuel, paper, and boat building increases. This practice has also led to a myriad of attendant problems such as desertification, excessive erosion and destruction of the watershed.

It would seem obvious that the innovators of the newer technologies concentrated on the supply end of the continuum with little or no built-in environmental safeguards for the users. On the other hand, industry, which is a user of technology is not always willing to take responsibility for environmentally unfriendly activities. In other words, risk assessment is not always factored into the technological equation as far as the interest of society is concerned. Modern-day society has placed a premium on the consumption of material goods. Unfortunately, the environmental cost of this kind of behaviour is not always immediately visible. To produce a single good, there is a long chain of processes that should also be accounted for when considering the environmental cost.

Environmental and sustainable manufacturing has not been treated as a matter of priority in many small island states. Over the last couple of decades, there have been vigorous sensitisation programmes highlighting a range of critical environmental issues as well as the sustainability of natural resources in vulnerable small island states. The apparent neglect to treat environmentally sustainable manufacturing as a matter of urgency is partly because most of these small states do not think that they are major polluters of the environment. This is because entities within these states are not renowned for heavy manufacturing activities that give rise to the production of heavy industrial waste and pollution.

The state has a responsibility to develop policies and guidelines and enact the necessary legislation to guide manufacturing enterprises. This will sensitise them about the fundamentals of environmentally safe and sustainable manufacturing practices. that will safeguard the health and safety of all citizens. As early as 1998, UNEP cited in a report on "Management of Wastes in SIDS" which stated that the lack of adequate treatment of solid wastes in many cities of small states including industrial wastes remains a major problem to be resolved. If good manufacturing practices are rigorously employed in the major manufacturing industries, then the problem would not transcend to one of containment and management of wastes. Not only should there be concerns about technologies employed that would increase productivity, but the issue of environmental safety and sustainability should also be considered when making decisions of a technological nature. Existing environmental standards that are compatible with international best practices for specific industries should be vigorously enforced among small enterprises. This will promote the use of environmentally sound technologies in industrial processes that will minimise or eliminate the incidence of industrial mishaps that could have detrimental impact on the environment.

There should be ongoing awareness among industries on the various issues that are associated with environmentally safe industrial practices. These technologies should be reengineered and made appropriate to the country and the enterprises using them, thereby ensuring that they comply with sound environmental standards that are designed to promote environmental safety. Manufacturing technologies should be assessed before they are procured and all limitations including limited resources, lack of technical skills and human resource capability of the countries to which they are being transferred should be considered so that the technology employed should not result in detrimental environmental effects which the country must cope with for a long time.

Methodologies, management systems and practices that will effectively control toxic chemical wastes emanating from advanced industrial processes should be developed. There should be documented procedural guidelines on the use and disposal of industrial wastes such as organic and inorganic wastes, heavy metals, contaminated liquids and sledges generated by SMEs through their regular simplified processes. Several steps can be taken to make manufacturing in any country more sustainable. Among the most effective steps are:

- Optimisation of the use of available renewable energy technologies
- Elimination of waste
- Reduction/elimination of pollutants
- Recycling of waste materials
- Recovery of energy.

Technologies to accomplish these measures are available on the market and are desirable and well proven. Some governments encourage the use of clean technologies by providing rebates to companies that procure and use them. For instance, renewable energy technologies which can reduce and eliminate pollution, waste reduction, recycling and energy recovery are driven by policy positions of companies supported by appropriate technologies.

Sustainable manufacturing approaches can also be harmonised and used effectively on a regional basis. It will require long-term planning as well as a coordinated effort between the private and public sectors to ensure that environmental standards set are adhered to. In addition, any technology proposed as clean technology should be considered safe for the health and well-being of the people who will be most affected.

References

Ajah, Jankey; Sipa, George, International Patents and transfer to Less Developed Countries, Gower Publishing Co. Ltd., 1987.

Anderson, David, R. (et al) An Introduction to Management Science: Quantitative Approaches to Decision Making, Fifth Edition, West Publishing Co. 1988, p.1

Anderson, M. G., & Katz, P. B. (1998). Strategic Sourcing. International Journal of Logistics Management, [cited 1998 January] 9 (1), 1-13.

Armstrong, H.W., Read, R.A. Trade and Growth in Small States: The Impact of Global Trade Liberalisation. Working Papers. Departments of Economics. Lancaster University,1998

Atlanta GmbH, Banana Fibre Processing, 1975.

Balassa, B. Trade Liberalisation and "Revealed" Comparative Advantage, The Manchester School of Economic and Social Studies, 1965.

Balassa, B. Comparative Advantage, Trade Policy and Economic Development. Harvester Wheatsheaf, New York. (1989).

Balasubramanyam, V.N. International Transfer of Technology to India, Praeger Publishers Inc., 1973.

Baldacchino, G. "A Tast of Small-Island Success: A Case from Prince Edward Island." Journal of Small Business Management [cited2002 December] 40(3) (2002): 254-259. Available from: https://doi[10.1111/1540-627X.00055](10.1111/1540-627X.00055)

Baranson, Jack; Industrial technologies for Developing Economies, Frederick Praeger Publishers, 1969.

Kav Michael Ying-Mao, Simon Denis Fred TAIWAN: Beyond the Economic Miracle, M.E Sharp Inc., Armonk NY, London, 1992, pp.124-125.

Beamon, M. B. (1998). Supply Chain Design and Analysis: Models and Methods. International Journal of Production Economics, 55, 281 – 294.

Behrman, J. Fischer, W.A. Science and Technology for Development, Cambridge, Gunn and Hain Publishers Inc., 1980.

Bell, Federick; Canterbury, Ray, Agriculture for the Developing Countries, Ballinger Publishing Company, 1976.

Bell, Federick; Canterbury, Ray, Aquaculture for the Developing Countries, Ballinger Publishing Co. 1976.

Bendall, Tony (et al). The Benchmarking Workout, Pitman Publishing, 1997. p.5, pp 30-31

Black, J.T. The Design of the Factory with a Future, Mc Graw-Hill Inc., 1991, p.15.

Bradley, Frank, Marketing Management: Providing Communication and Delivery Value, Prentice Hall, 1995.

Burbidge, John. L Group Technology in the Engineering Industry, Mechanical Engineering Publications Ltd. London, 1979, pp. 106-107.

Caribbean Community; Caribbean Trade & Investment Report 2005: Corporate Integration & Cross-border Development; Caribbean Community Secretariat, 2006.

Caribbean Trade & Investment Report 2005.

Chan, F. T. S., & Chan, H. K. (2006). A Simulation Study with Quantity Flexibility in a Supply Chain Subjected to Uncertainties, International Journal of Computer Integrated Manufacturing, Vol. 19, No. 2, 148 – 160.

Chandra, R. Industrialization and Development in the Third World, Routledge, 1992, pp. 50-51.

Chandra, Rajesh, Industrialization and Development in the Third World, Routledge, 1984.

Chase, Richard B.; Aguilano, Nicholas J. Production and Operation Management: Manufacturing Services, Seventh Edition Richard D. Irwin. Inc. 1995, p. 413, 609

Competitiveness and the Role of ICT in the Caribbean Copyright infoDev 2005 – www.infodev.org

Cook Paul, (et al) Privatization, Enterprise Development and Economic Reform: Experiences of Developing Transitional Economies, Edward Elger Publishing Ltd. 1998, p. 216.

Cusumano, M.A. The Japanese Automobile Industry, Harvard University Press, 1989.

de la Mare, R.F. Manufacturing Systems Economics, Holt, Rinehart and Winston Ltd., 1982, p. 5.

Department of Trade and Industry White Paper (1994), Competitiveness: Helping others to win." Pp136-143.

Derakhshani, S. Factors Affecting Success in the International Transfer of Technology: A Synthesis and a Test of a New Contingency Model: Developing Economies 21, 1982.

Dimancescu, Dan; Dwenger, Kempt, World Class New Product Development: Benchmarking Best Practice of Agile Manufacturing, Amacon, American Management Association, 1996, p. 21.

Dixon, C. South East Asia in the World Economy: A Regional Geography, Cambridge University Press, 1991, p.162.

Dodgson, Mark; Rothwell, Roy, The Handbook of Industrial Innovation, Edward Elgar Publishing Company, 1994.

Dunn, Peter, Appropriate Technology: with a Human Face, McMillan Press Ltd., 1978.

Dunning, J.H., "Alliance, Capitalism and Global Business", London, Routledge, 1997.

Durand, Martin and Giorno, Claude, Indicators of International Competitiveness: Conceptual Aspects and Evaluation - OECD Economic Studies No. 9, Autumn 1987 pg. 149.

ECLAC International Trade Outlook for Latin America and the Caribbean 'Stronger regional integration urgent to counter impact of trade conflicts." [cited 2018 December]. Available from https://doi: https://hdl.handle.net/11362/44197.

ECLAC "Efficacy of Caribbean Industrial Policy in the New Era of Globalization." [cited 2001 June] Available from: https://www.cepal.org/sites/default/files/publication/files/38434/ pdf.

ECLAC GENERAL LC/CAR/G.576, "Competitiveness of the Manufacturing and Agro-Industrial Sectors in the Caribbean with a

Focus on Dominica, Guyana, Saint Vincent and the Grenadines and Trinidad and Tobago," 15 November 1999.

Ernst, Dieter, O'Connor, David, Technology and Global Competition: The Challenge of Newly Industrializing Economies, OECD, 1989.

European Commission, A New Industrial Strategy for Europe, Brussels, March 2020.

Evans, James, R. Linsdsay; William, M. The Management and Control of Quality, Fourth Edition, South-Western College Publishing, 1999, p.120.

Feigenbaum, A.V. Total Quality Control: Engineering and Management, Mc Graw-Hill, 1961, p.12

Fiji Ministry of Finance Central Planning Office 1990. Available at: http://worldcat.org/identities/lccn-n50005068/.

Focus on Fibres, New Agricultural On- Line 99-4. Available from: http://www.new-ag.info/99-4/focuson.html.

Freytag, Per V.; Hollensen, Svend, The Process of Benchmarking, Bench learning and Bench action", Total Quality Magazine, Vol. 13. 2000, p.2.

Gee, Sherman, Technology Transfer, Innovation and International Competitiveness, John Wiley and Sons, 1991.

Giovanna Ceglie and Marco Din (UNIDO consultant) "SME Cluster and Network Development in Developing Countries": The Experience of UNIDO, 1999.

Girvan, Norman, "Technological Change and the Caribbean: Formulating Strategic Responses: Social and Economic Studies," Vol. 2, Sir Arthur Lewis Institute of Social and Economic Studies, 1989.

Global Forum on Industry: "Perspective for 2000 and Beyond," UNIDO, 1995, pp. 2-3

Globe, et.al, Innovation: The Creation and Implementation of New Technology, Tappi, 1974.

Gray, Jerry; Starke Frederick, Organizational Behaviour, Charles E. Merrill Publishing Company, 1980, p. 125&339.

Gregory, S, A., Creativity and Innovation in Engineering, National Significance of Technology, Newnes-Butterworth (October 1, 1972.

Gunn, Thomas G. Computer Applications in Manufacturing", Industrial Press Inc. 1981, p. 174

Hall, Robert, W. Attaining Manufacturing Excellent, Dowe Jones-Irvin, 1987. pp. 41-42

Hammervoll (2009).

Handel, M. 1981. Weak State sin the International System. New York: Frank Cass & Co. Ltd.

Hannam, R.G. Kaizen for Europe: Customising Japanese Strategies for Success, IFS Ltd. U.K. 1993, p.48.

Harold, Chec; Rod, Hariis, Marketing: A Global Perspective, Pitman Publishing, 1993.

Harrington James, Benchmarking with Dr. James Harrington, Learner First, Inc. 1993.

Harrison, Alan, Just-in-Time Manufacturing in Perspective, Prentice Hall International, 1992.

Hazeltine, Baret, Appropriate Technology: Tools, Choices and Implications, Academic Press, 1999.

Hendry, C. (et al.) Human Resource Development in Small to Medium Size Enterprises (Research Paper No. 88), Shellod Employment Department, 1991.

Hollingum, Jack, Implementing an Information Strategy in Manufacture: A Practical Approach, IFS (Publications) Ltd U.K., 1987, p. 4-5.

Holt, Knut, Product Innovation Management, Second Edition, Butterworth, 1983.

House of Lords Select Committee on Science and Technology, Third Report: Clusters, Science Parks, Incubators and Management Education, 1997.

Hoyle, David, ISO 9000: Quality System Handbook, Butterworth-Heinemann, Third Edition, 1998, pp. 22-23.

Huff, W.G, The Economic Growth of Singapore: Trade and Development in the Twentieth Century, Cambridge University Press, 1994.

Huq, Mozammel [et.al] Science, Technology and Development: North-South Cooperation, Frank Cass and Co. Ltd. 1991.

Huq, Mozammel [et.al] Science, Technology and Development: North-South Cooperation, in Hashem and Ali, Haerian; Manpower Constraints in the Industrial Recovery of Post War Iran op cit. Frank Cass and Co. Ltd. 1991.

Huq, Mozammel [et.al] Science, Technology and Development: North-South Cooperation, in Yoaxing Hu; Science and Technology in China, op cit. Frank Cass and Co. Ltd. 1991.

Hutchins, David, Quality Circles Handbook, Pitman Publishing Ltd., 1985, pp 8-9.

Imai, Masaaki, Kaizen: The key to Japan's Competitive Success, Mc Braw-Hill Publishing Company, 1986, pp.3,6.

IMF Working Paper, Prepared by Sebastian Acevedo. Gone with the Wind: Estimating Hurricane Climate Change Costs in the Caribbean, October 2016.

Improving Competitiveness and Increasing Economic Diversification in the Caribbean: The Role of ICT, April 2005.

Improving Competitiveness and Increasing Economic Diversification in the Caribbean. Available from: https://www.infodev.org/infodev-files/resource/InfodevDocuments_5.pdf.

J, Luis, Gausch, Jean-Louis, Racine, Isabel Sanchez and Makhtar Diop; Quality Systems and Standards for a Competitive Edge, World Bank, 2007.

Jobber, David, Principles and Practice of Marketing, Second Edition, Mc Graw Hill Publishing, 1998.

Joseph Prokopenko, "Productivity Management; A Practical Handbook" International Labour Office, 1987.

Karlof Bengt, Benchmarking Workbook: With Examples and Ready-Made Forms, John Wiley and Sons Limited, 1995. p. 5.

Kartick K. Samanta and S.N. Chattopadhyay, Properties, Processing and Application of Banana Fibre International Journal of Bioresource Science [cited 2022 December] 9 (2); Available from: https://doi DOI:10.30954/2347-9655.02.2022.15

Kav Michael Ying-Mao, Simon Denis Fred TAIWAN: Beyond the Economic Miracle, M.E Sharp Inc., Armonk NY, London, 1992, pp.124-125.

Khalid Nadvi & Frank Wältring. Making Sense of Global Standards, December 200.1

Kiggundu, Mosas, Managing Organizations in Developing Countries: An Operational and Strategic Approac, Kumarian Press Inc. 1989.

Kirton, C. and Tenant, D. (2008) Policies and Institutions Supporting Small and Medium Scale Enterprises (SMEs) in Jamaica.

Klapper, Norman, D. HI/MIS High Impact Manufacturing Systems, Van Nostrant Reinhold, 1992.

Kotler, Philip; Armstrong, Gary, Principles of Marketing, Prentice- Hall International, Fifth Edition, 1991.

Kulessa, Manfred, The Newly Industrializing Economies of Asia: Prospects for Corporation, New York 1990, Springer Verlog, 1993, pp. 97&105.

L. Gray Cowan, Privateization in the Developing World, Praeger Publishers, 1990.

Lancaster, G. Massingham, L. Essentials of Marketing, Second Edition, Mc Graw Hill Book Co., 1993.

Le Sann, Allan; A Livelihood for Fishing, Intermediate Technology Publications, 1998.

Lee, Fred, Project for Transforming and Valorising Banana plant Waste, (Paper presented at the workshop on the alternative use of banana) in St. Lucia.

Leenders, M. R., Johnson, P. F., Flynn, A. E., & Fearson, H. E. (2006). Purchasing and Supply Management. New York: McGraw-Hill/Irwin.

Lewis-Bynoe, D, J Griffith, and W Moore. Trade Liberalisation and the Manufacturing Sector: The Case of the Small Developing Country. Contemporary Economic Policy 20, no. 3 (2002): 272-287.

Likert, Renis, The Human Organization, MC Graw-Hill 1967.

Lockhart, Douglas [et al] The Development Process of Small Island States, Routledge, 1993, 16, 40; pp.210-222

Lou, Cohen, Quality Function Deployment: How to make QFD Work for You, Addison-Wesley Publishing Company, Inc., 1995.

Luscombe, Martyn "MRP II Integrating the Business: A Practical Guide for Managers, Butterworth Heinemann, 1983.

Madeley, John, Trade and the Poor: The Impact of International Trade on Developing Countries, Intermediate Technology Publications Limited. 1992.

Malhotra N.K. Marketing Research: An Applied Orientation, Third Edition, Prentice Hall Inc., 1998.

Marsden, K, Progressive Technologies for Developing Countries, Small Industries Unit, International Labour Organization, 1967.

Mc Intyre, Arnold, "Science and Technology Policy in Developing Countries: Some Implications for the Commonwealth Caribbean", Yale University, 1987.

Meissner, Frank; Technology Transfer in the Developing World, Praeger Publishers Inc., 1988.

Menon, M.G. [et al] Commonwealth Secretariat, Report by a Commonwealth Working Group: Technological Change: Enhancing the Benefits, Vol.1., 1985

Mercer, David; Marketing, Blackwell Publishers, 1992.

Miles, Derek, Constructive Change, International Labour Organization, 1995.

Miller, Landon, Concurrent Engineering Design: Integrating the Best Practices for Process Improvement, Society of Manufacturing Engineers, 1993.

Molander, Christopher, Human Research Management, Chartwell Bratt Ltd. 1989.

Mondem, Yasuhiro, Toyota Production System: An Integrated Approach to Just-in-Time, Second Edition, Chapman & Hall 1994, p.16.

Morden, Anthony, Elements of Marketing, Second Edition, 1991, The Guensey Press Co. Ltd.

Morris, Alan, S. Measurement and Calibration Requirements for Quality Assurance to ISO 9000, John Wiley& Sons, 1999, p.19, pp.34-37.

Nadu, Christian, Strategic Planning in Technology Transfer for Less Developed Countries, Qurum Books, 1992.

Nakajima, Seiichi Introduction to TPM, Productivity Press, Portland OH. 1988.

National Economic Development Council (NEDC), Innovation Working Party on: Technology Transfer Mechanisms in the U.K and leading Competitor Nations, National Development Office, 1989.

Nelson Correa (UNIDO) Foteini Kanatsouli (UNIDO Consultant) Industrial Development in Least Developed Countries, UNIDO Working Paper 26, 2018.

Oakland, John, I. Statistical Process Control, Fourth Edition, Butterworth Heinemann, 1999, pp. 15-16.

OECD (1971), Conditions for Success in Technological Innovation. Issue 28 of OECD publication Organisation for Economic Co-operation and Development, 1971.

OECD, Technology Incubators: Nurturing Small Firms. OECD 1997.

David Rabkin, Marcela Escobari, Camila Rodriguez OTF Group. 2005. Improving Competitiveness and Increasing Economic Diversification in the Caribbean: The Role of Information and Communication Technologies. Published 1 April 2005.

Pacific Economic Monitor, December 2016. Available from: https://www.adb.org/sites/default/files/publication/189886/pem-dec-2016.pdf.

Pebbles Gavin; Wilson Peter, The Singapore Economy, Edward Elgar Publishing Ltd. 1996, pp. 1-4.

Hines Peter, Creating World Class Suppliers: Unlocking Mutual Competitive", Financial Times, Pitman Publishing, 1994. p.200.

Porter M.E, The Competitive Advantage of Nations; Harvard Business Review, 1990.

Productivity Digest, Learning from the Best Practices, March 2001 Vol 20 No1.

Proposals on Investment Policy Harmonisation and Coordination in the Caribbean Community, March 2004.

Pye, Lucian, W. In Search of an Asian Development Model, Paper presented at a conference sponsored by the Carnegie Council on Ethnics and m International Affairs, held in New York, 1978. p. 79.

Ranson, G.M. Group Technology, Mc Graw-Hill, 1972, p.2

Rawlinson, Greffery, Creative and Brainstorming, Gower Publishing Co. 1981.

Report of ILO Technical Support Services (TSSI) Mission: Small Business: Key Ingredients and Constraints to their Success in the Caribbean," 1995.

Robson, Mike, Quality Circles: A Practical Guide, Gower Publishing Co., 1982, p.3.

Roelandt, T. den, Hertog, P. Summary Report of the Focus Group on Clusters, 1997.

Rosenberg, Nathan; Frischtak, Cladio, International Technology Transfer: Concepts, Measures and Comparisons, Praeger Publishers, 1985.

Rosenburg, Nathan; Frischtak, Cladio, International Technology Transfer: Concepts, Measures, and Comparisons, Praeger Publishers, 1988.

Sahal, Devendra, The Transfer and Utilization of Technical Knowledge, Lexington Books, 1990, p. 110

Sahal, Devendra, The Transfer Utilization of Technical Knowledge, Lexington Books, 1990.

Samli, Coskun, Technology Transfer, Geographic, Economic, Cultural and Technological Dimensions, Quorum Books, 1985.

Schonberger, Richard, J. World Class Manufacturing: The Lessons of Simplicity, The Free Press, 1986, p.192, 217.

Schonberger, Richard, J. World Class Manufacturing: The Next Decade, The Free Press 1996, pp. 5,6,9, 16.

Science and Technology and Development, in Gross-Hertog Irene Recent Experiences with Appropriate Technology and Implications for Technical Vocational Training, op cit. Vol. 3, Frank Cass, 1998.

Science and Technology and Development, in Magabe, J. and Clark N.: Technology Transfer and the Biodiversity Convention op cit. Frank Cass.

Science Direct; ELSEVIER; Innovation and Technology Transfer for Business Development. Available from: https://doi.org/10.1016/j.proeng.2016.06.697.

Science, Technology and Development, in Adeboye Titus, "Technological Capabilities in Small and Medium Enterprise Clusters: Review of International Experience and Implication for Developing Countries, op cit. Frank Cass Vol. 14, 1996.

Sebastian James. Incentives and Investments; Evidence and Policy Implications, World Bank Group, 2009.

See Arthur Lewis, The Industrialization of the British West Indies, Caribbean Economic Review, 1950. U.S. Government Printing Office, 1950.

Simon, Dennis (1992) Taiwan, Beyond the Economic Miracle, East Gate Books. p.4, pp. 124-125.

Singh, Nannon; Rajamani, Divakar, Cellular Manufacturing Systems Design, Planning and Control, Chapman & Hall, 1996.

Skinner, W. (1969). Manufacturing: Missing Link in Corporate Strategy. Harvard Business Review, May-June, 136-145.

Habolm. Sarwar, Small Medium Sized Enterprises in Economic Development; The UNIDO Experience, 2008.

Smith, Robert (Graduate Research) A Guide for Students in the Sciences, University of Washington Press, 1998.

Sohal, Amrik; Ritter, Mark, Manufacturing Best Practices; Observations from study tours to Japan, South Korea, Singapore and Taiwan, Benchmarking for Quality Management and Technology, Vol. 2, Issue 4, 1995.

Soon, T. and W. A, Stoever, "Foreign Investment and Economic Development in Singapore:

Sorak, M. & Dragic, M. Supply Chain Management of Small and Medium-Sized Enterprises, DAAAM International Scientific Book 2013 pp. 951-968.

Stabaugh, Wells, Technology Crossing Borders, Harvard Business Press, 1984.

Stephenson 1997, Sengenberger and Campbell 1994, OECD 1995, 1996,

Stewart (1987), Handbook on Appropriate Technology, 1976.

Stewart, Francis; Macro Policies for Appropriate Technology in Developing Countries, Westview Press Inc. 1987.

Suzana N. Russell1 & Harvey H. Millar; Competitive Priorities of Manufacturing Firms in the Caribbean. IOSR Journal of Business and Management (IOSR-JBM) Volume 16, Issue 10.Ver. I (Oct. 2014), PP 72-82.

Syan, S. Chanan and Menon Unny, Concurrent Engineering: Concepts, Implementation and Practice, Chapman & Hall 1994.

Technological and Human Recourse Development, 2002. Available from: https://hdl.handle.net/11362/27523.

Technological Change: Enhancing the Benefits Volume 2: Report of a Commonwealth Working Group, Commonwealth Secretariat, 1985.

The Computer and Automated Systems, Association of SME (selected articles): "Capabilities of Group Technology, 1987.

The Courier, The Magazine of ACP-EU Development Cooperation, Dossier, Least Developed Countries, p13.

The Courier: ACP-EU, Third UN Conference, Directorate General for Development, EC. 2001.

The House of Lords Select Committee on Science and Technology, Third Report (1991). Available from: https://publications.parliament.uk/pa/ld199900/ldselect/ldsctech/38/3803.htm.

The World Bank Group; "Private Sector and Infrastructure Network." February 2003. http://rru.worldbank.org/Viewpoint/index.asp

The World Bank Research Observer; [Cited 1996 August] (11) 2; 151-177. Available from: https://doi.org/10.1093/wbro/11.2.151

The World Bank, Time to Choose; Caribbean Development in the Twenty First Century, April 2005.

The World Economic Forum's Global Competitiveness Report 2007 – 2008. Available from: https://www3.weforum.org/docs/WEF GlobalCompetitivenessReport 2008-09.pdf.

Todaro, Michael, Economic Development in the Third World, Fourth Edition, Longman Publishing, N.Y., 1989.

Todd, Jim, World-Class Manufacturing, Graw-Hill Book Company, 1995.

Touminen, Kari, Managing Change: Practical Strategies for Competitive advantage, American Society for Quality, 2000.

Twist, Brian, Managing Technological Innovation, Fourth edition, Pitman Publishing, 1992.

UNCTAD (1993), Draft International Code of Conduct on the Transfer of Technology Available at: https://core.ac.uk/download/pdf/216913331.pdf

UNDP, Max Everest-Phillips; "Small, So Simple? Complexity in Small Island Developing States." 2014 UNDP Global Centre for Public Service Excellence.

UNECLAC, Building SME Competitiveness in the European Union and Latin America and the Caribbean, January 2013. Available at: https://www.cepal.org/en/publications/3093-building-sme-competitiveness-european-union-and-latin-america-and-caribbean-policy.

UNECLAC 'Impact of New technology on the Development Process in the Caribbean, Jan. 2003. Available at: https://www.cepal.org/en/publications/27524-impact-new-technologies-development-process-caribbean-region.

UNECLAC Impact of New Technology on the Development Process in the Caribbean, Jan. 2003. Available from: https://hdl.handle.net/11362/27524.

UNECLAC/ CDCC. 2000, Industrialization, New Technologies and Competitiveness in the Caribbean. Available from: https://hdl.handle.net/11362/27462.

UNECLAC/CDCC, Adoption and Application of Information Technology in the Caribbean and its contribution to Scientific, Technological and Human Recourse Development, 2002. Available from: https://hdl.handle.net/11362/27523.

UNECLAC/CDCC, Impact of New Technologies on the Development Process in the Caribbean Region, 2003. Available from: https://hdl.handle.net/11362/27524

UNIDO Industrial Development in Least Developed Countries, Working Paper 5/2012.

UNIDO, "A Manual on Technology Transfer Negotiations A reference for policy-makers and practitioners on technology transfer (General Study Series), 1996.

UNIDO, Amadou Boly, Industrial Development in Least Developed Countries, Statistics Unit Strategic Research, Quality Assurance and Advocacy Division UNIDO, 2013.

UNIDO: Industrial Development Report 2009. "Breaking In and Moving Up: New Industrial Challenges for the Bottom Billion and the Middle-Income Countries." Available from: https://www.unido.org/sites/default/files/2009-02/IDR_SUMMARY_print_0.pdf.

Urban, Glen; Hauser, John, Design and Marketing of New Products, Second Edition, Prentice Hall Inc. 1993.

Ventura Arnaldo, Science and Technology for Poverty Alleviation and Social Development in the American Hemisphere, 1996.

Virmani, B.R, Rao kala, Economic Restructuring, Technology Transfer and Human Resource Development Response, Sage Publication Inc., 1997, p. 32.

Vlaadimir, Ignatieff, A Handbook on Appropriate Technology, Education Committee, Canadian Hunger Foundation, Ottawa, 1976.

Vollmann, Thomas, E. Manufacturing Planning and Control Systems Second Edition Richard D Irwin Inc, 1988.

Von, Weizsaker, E.N., New Frontiers in Technology Application, in Lydia Makhubu: "Integration of Emerging and Traditional Technologies", op cit International Publishing Ltd. Vol. 2. 1983.

Wallace, Thomas, F. MRP II: Making it Happen: The Implementers' Guide to Success with Manufacturing Resource Planning, John Wiley & Sons, Inc. 1990, pp.5, 11.

Weiss, John Industry in Developing Countries, Theory, Policy and Evidence, Croom Helm Ltd., 1988, pp.28-29.

Willmore, L. Export Processing in the Caribbean: Lessons from Four Case Studies. Working Paper, United Nations, Santiago: Economic Commission for Latin America and the Caribbean, 1996. Available at: https://ideas.repec.org/p/wpa/wuwpit/0412001.html.

Zaleski, E.; Wienert H. Technology Transfer between East and West, Organization for Economic Co-operation and Development; Washington, D.C. 1980.

Index

A

A Model, 233
A model of business support, 111
A philosophy of change, 113, 302
A product, 75
A standard specification, 164
A strategy, 81
A tax incentive, 41
A world-class approach, 137
A world-class manufacturing
    approach, 134
Access to broadband services, 119
Achievement of measurable
    benchmarks, 289
Achieving world-class manufacturing
    status, 134
African Small Island Developing
    States, 23
Aid for Trade, 13, 14, 13–14
Air freight, 292, 294
An
    incubator, 106
    industrial park, 46
    industrial policy, 251
    organization, 75
appropriate technology, 67, 162
Asian Development Bank, 39, 293
Asian, Pacific Island States, 25

B

Backward integration, 94 *See also*
    Forward and backward linkages
Barriers to the growth of SMEs, 103
Benchmarking, 160–62, 203
Best practices in port operations, 293
Bioprospecting, 284
Brainstorming, 78
brand name, 181, 281
Branding and, 283
Breakthrough innovations, 88
Building technological capacity,
    52, 242
Business
    incubation, 303
    incubators, 48, 101, 107, 110
    support programme, 112

C

Capital accumulation, 243, 268
Capital and technology-intensive,
    235, 238
Caribbean Community, 3, 269
challenges of smallness and
    remoteness, 24
challenges posed by the bureaucratic
    red tape, 304
Cluster of firms, 225

Clustering of small enterprises, 103
Clusters are also defined, 102
Clusters in small developing state, 106
Commercial innovation, 83, 112,
Common External Tarif, 4, 259
Comoros, 24
Comparative advantage in service delivery, 249
compete directly in the global market, 50
competitive advantage in export trade, 29
Competitive Advantage of Nations, 175
Competitive bidding, 296
Competitive Priorities of Manufacturing Firms, 218
Competitiveness and Innovation, 276
Competitiveness in the global economy, 181
Concurrent Engineering, 140–141
Concurrent or 'simultaneous' engineering, 140
Continuous improvement, 141
Cooperation through business association, 285
Corruption in governance, 310
Cost of poor governance, 310
Creativity, 77–79, 96
Creativity and innovation, 247
Cross-border dispersion, 307
Cross-border e-commerce, 117
Cross-border E-commerce, 121
Customer delight, 150
Customer satisfaction, 89, 90
Cyber Parks, 47, 48

D

Definition of smallstates, 19
Definitions of innovation, 75, 76
De-industrialization, 8, 33, 248
Demand side of the market, 254

Developing country, 12
Development challenges, 16, 28
Development of a quality workforce, 245
Digital
  economies, 117
  ecosystem, 117
  technologies, 117, 246
  transition, 122, 246
Diversification of the agricultural sector, 7
Diversify away from manufacturing, 36
Duty-free imports, 43, 45

E

Economic
  and fiscal policies, 240
  diversification, 26, 28
  independence, 209, 264
  diversification programmes, 124
  integration, 3
Education and training, 65, 273
Education is not limited to school and university, 273
Education is the basis for change, 272
Effective incentive scheme, 38, 254, 290
Effective supply chain management system, 181
effectiveness of fiscal incentives in attracting investment, 290
efficient intra-regional transport service, 293
Efficient sea and air transport service, 292
Electronic commerce, 115
Electronic Transactions Act, 120
Enclave industries, 106, 233, 235
Endowment and peculiarities of small states, 261

Energy as a tool to promote sustainable development, 300
Energy security, 301
Enterprise Resource Planning, 123, 126
enterprise/ incubation development centres, 53
enterprises from indigenous commodities, 226
Entrepreneurial culture, 274
Entrepreneurship, 274–275
environmentally safe industrial practices, 316
Environmentally sustainable Manufacturing, 315
escallion, 94
Establish a regulatory body, 301
Establishing ICT related business, 119
Establishment of a network of support, 298
Export Processing Zones, 44, 265
Export-led growth, 233, 235
Export-led industrialization, 242
Export-oriented industrialization model, 248

F

Filtronic plc, 275
Financial incentives, 36-37, 43, 220
financing to these non-traditional sectors, 288
Firms in a cluster, 101
Fiscal incentives, 36
Fiscal subsidies, 291
Foreign Direct Investment, 40, 43, 278
Formation of a public private partnership, 297
Forward and backward linkages, 10, 34, 271
Foster and promote innovation, 83
Free Trade Zones (FTZ), 43

G

General Sun Tzu, 160
Global markets, 164, 225
Globalization of production, 9, 11, 172
Good Agricultural Practices, 187
Good industrial relations, 309
Governments can play a critical role, 243
Green environment, 313
Guinea-Bissau, 24
Gut Feel is Market Research, 92

H

Harmonized investment policy framework, 289
Harmonized policies, 257
Heckscher-Olin theory, 215
Hertzberg, 138

I

Impacted by global economic crises, 30
Import Substitution Industrialization, 234
Import substitution policy, 11
In the global economy, 178
increasing carbon emissions, 313
incubators in the Caribbean, 111, 303
Infrastructure provisions, 286
Infrastructure to facilitate the transit of goods, 287
Innovations in management, 133
institutional capacity and regulations, 287
Institutional strengthening, 306
Intellectual property, 280
Intellectual property law, 283
International
    competitiveness, 213
    standard bodies, 153

Standardisation Organization, 164
standards of excellence in quality, 133
Internationalization of production, 1
Investment allowances, 36, 38, 41, 42
Investment in information communication technology, 125
Investment tax credits, 41
Involvement and contribution of its employees, 158
Inward-looking policy, 238 *See also* import substitution industrialization
IP provides the requisite protection, 283
Irreversible impact of climate change, 312

K

Kaizen, 141
Kaizen philosophy, 141
Key inputs to the success of any company, 80

L

Lack of private sector involvement, 27
lack of well developed capital markets, 244
Large and medium-sized enterprises, 34
Laws on intellectual property, 126
Least Developed Countries, 9, 336
leather shoemakers, 68
Lending institutions in small states, 244

M

Management by Objectives, 138
Management techniques and philosophies, 139
Manufacturing, 32, 221

Environmentally sustainable, 315
Lean Manufacturing, 142
Market efficiency, 296
Market failure, 11, 245, 252
Market Pull, 82
Marketing
  agency, 308
  and distribution agencies, 308
  concept, 88
  research, 92
Mauritius, 24, 126, 239
Modern marketing, 84
monopolistic status, 295

N

Narrow the technology gap, 243
National Innovation Systems, 97
national manpower planning, 53, 276
Nauru, 21, 24
neo-classical hypothesis, 210
Networking among firms, 304
Networking promotion agencies, 115
Networks among small firms, 115
New and emerging energy technologies, 299
New International Economic Order, 5
New product development, 84-85
Non-commercial channels, 64

O

Odiorne and Humble, 138
Organization of East Caribbean States, 23
Organization of Eastern Caribbean States, 3
Outsourcing or subcontracting arrangements, 298
Outward-looking economy, 258 *See also* Export led industrialization

P

Pacific Island Countries, 3, 26, 293
Pacific Islands Forum, 3, 25
Papua New Guinea, 24
participatory management techniques, 154
Patent and Trademark Office, 282
Performance measurement, 200
Performance-based specifications, 165
Policies require continuous monitoring, 262
PPP model, 296
Preventive maintenance, 145
Principles of effective marketing, 86
private sector involvement, 168, 296
Privatization, 209, 211
Produce Chemist Laboratories, 184, 231
Product innovation, 77, 85
Production and consumption patterns, 312
Production integration, 307
of a Caribbean Steelpan, 280
planning and control, 140
Productivity, 195, 199
productivity measurements, 200
Protectionist policies, 245, 256
Public-Private Partnership, 294

Q

quality assurance, 151-153
Quality Assurance, 151–53
Quality circles, 138, 154
quality for cost, 95

R

Rationale for the development of an industrial policy, 253
Reactive product strategy, 93
Reduce their carbon footprint, 313
Reducing liability risk, 296
Regional integration, 226, 307
Regional integration agreements, 3
Regionalism, 2, 3

S

Science and Technology, 96
Sectoral approach to clustering, 102, 103
shift away from agriculture, 238
Single market and economy, 3
Single point contact or electronic window, 305
Six Sigma, 158
SM Jaleel, 80
Small and micro enterprises, 34, 112
Small Island Developing States, 9
Small size and remoteness, 26
Small states., 274–275
Smallness, 18, 27
Sole proprietorship, 35
Sophisticated technological infusion, 305
South-South Cooperation, 13–14
Standard Operating Procedures, 153
Statistical Process Control, 155
Strategic marketing analysis, 93
Structural transformation, 7 *See also* Complementarity
Sub-grouping umbrella organization, 23
Successful transfer of technology, 55
Suggestion System, 79
Supplier-customer relationships, 183
Supply-side oriented policies, 255
Supply-side policies, 260
Sustainable energy, 300
Sustainable industrial climate, 311

T

Targeted agricultural production, 25
Tax allowances, 41
Tax holidays, 39, 40
Technical and Vocational Education and Training, 275
Technological
    capability, 54
    embodiment, 57
    innovation, 77
Technology
    extension services, 279
    incubators, 106
    technology innovation centres, 107
    transfer modalities, 59
The newly industrializing economies, 234
The complexity of an innovation process, 77
The concept of
    Business Incubation, 111
    Manufacturing, 129
    Quality, 149
The criteria for an SME, 99
The efficacy of investment incentives, 290
The global economy, 29, 52, 169, 174
The global value chain, 177
the globalization of industry, 6
The industrial achievements of the NICs, 239
The Industrial Revolution, 130
The knowledge economy, 247
The main element of economic success, 236
The philosophy of business incubation, 303
The process of innovation, 78
The process of innovation, 76–78
The production system, 129
The Sectoral Approach to Clustering, 102–103
The services sector, 11, 212, 248
The story of Qiaotou, 217
The Strategic Continuous Improvement Process, 193, 204
The term
    industrial parks, 45
    technology transfer, 51
    World-Class, 128
The word industry connotes, 32
the World Bank, 11, 290
Timor-Leste, 26
Total Productive Maintenance (TPM), 145
Total quality control, 157
Total Quality Management, 156, 159, 170
Trademarks and copyrighted works as economic assets, 284
Transfer of scientific or technological knowledge, 51, 276
Tuvalu, 21
Typical incubation system, 303
Typical tax incentives, 38

U

United Nations Industrial Development Organization, 6
United States Patent and Trademark Office, 280

V

Value-added, 221
Virtual shopping areas, 121
Vulnerability of small states to external shocks, 31

W

Walkerswood, 94
Weakness in their institutional capacity, 304
Website creation and operating services, 120
World Trade Organization (WTO), 255, 257
World-class Manufacturing techniques, 8, 137, 163 *Seee also* management techniques and philosophies
WTO agreement, 4

www.ingramcontent.com/pod-product-compliance
Lightning Source LLC
Chambersburg PA
CBHW020626220526
45464CB00001B/42